Technology Development Assistance for Agriculture

The DFID Research into Use Programme (RIU) has been a 6 year measure designed to scale out results from earlier DFID funded research. *Technology Development Assistance for Agriculture* reviews part of this programme as a case study of the broader issue of technology development for Africa. Controversially, it critiques current international technology development assistance and focuses on the potential role of the private sector in agricultural technology development, as well as providing insights for future science policy and practice.

The book focuses on the RIU Best Bets Africa sub-programme. This identified promising proposals to take existing agriculture research products and put these into use in ways that would benefit the poor in developing countries. The sum set aside for this was £5 million. The empirical sections of the book cover project selection, progress and programme management over the 2009–2012 period, with special attention paid to lessons learned that may have implications for future cognate technology development assistance.

This topical book gives direct evidence of meeting objectives and delivering real changes in technology development for Africa to postgraduate students, researchers, international bodies, NGOs, policy makers and government organisations working on natural resource management, technology development assistance and low income country agriculture.

Norman Clark is Professor of Innovation Systems and Development, Open University, and Senior Adviser to the DFID Research into Use Programme, UK.

Andy Frost is Deputy Director for DFID Research into Use Programme, University of Edinburgh, UK.

Ian Maudlin is an Honorary Professor, College of Medicine & Veterinary Medicine, University of Edinburgh and Director of DFID Research into Use Programme, UK.

Andrew Ward is a consultant based in Zambia and the UK.

Routledge Explorations in Development Studies

1. **The Domestic Politics of Foreign Aid**
   Erik Lundsgaarde

2. **Social Protection in Developing Countries**
   Reforming Systems
   Katja Bender, Markus Kaltenborn and Christian Pfleiderer

3. **Formal Peace and Informal War**
   Security and Development in the Congo
   Zoë Marriage

4. **Technology Development Assistance for Agriculture**
   Putting Research into Use in Low Income Countries
   Norman Clark, Andy Frost, Ian Maudlin and Andrew Ward

Technology Development Assistance for Agriculture

Putting research into use in low income countries

Norman Clark, Andy Frost, Ian Maudlin and Andrew Ward

Routledge
Taylor & Francis Group
LONDON AND NEW YORK

Research Into Use

First published 2013
by Routledge
2 Park Square, Milton Park, Abingdon, Oxon OX14 4RN

Simultaneously published in the USA and Canada
by Routledge
711 Third Avenue, New York, NY 10017

Routledge is an imprint of the Taylor & Francis Group, an informa business

British Library Cataloguing in Publication Data
A catalogue record for this book is available from the British Library

Library of Congress Cataloging-in-Publication Data
Technology development assistance for agriculture : putting research into use in low income countries / edited by Norman Clark ... [et al.].
p. cm. -- (Routledge explorations in development studies ; 4)
Includes bibliographical references.
1. Agricultural assistance--Developing countries. 2. Agricultural development projects--Developing countries. I. Clark, Norman. II. Series: Routledge explorations in development studies ; 4.
HD1431.T395 2013
338.1'8--dc23
2012045693

ISBN: 978-0-415-82697-6 (hbk)
ISBN: 978-0-415-82702-7 (pbk)
ISBN: 978-0-203-52489-3 (ebk)

Typeset in Sabon by
Saxon Graphics Ltd, Derby

MIX
Paper from responsible sources
FSC® C018575
www.fsc.org

Printed and bound in Great Britain by MPG Printgroup

'An inspirational source of ideas on how to translate research results into economic outputs. This book is an essential handbook for policy makers, practitioners and students of innovation.'

Professor Calestous Juma, Harvard Kennedy School, USA

'This book provides an excellent analysis of an attempt on the part of British overseas aid to deal with a major issue confronting the small farmer in low income countries: how best to translate good science into sustainable innovation and development. It should be essential reading for those involved in development policy.'

Professor Sir Gordon Conway, Imperial College London, UK

'This book probes the critical question of why the benefits of scientific knowledge have not been enjoyed equally by region and socio-economic class. Drawing from the Best Bets programme, innovative strategies of how to put research into use for the benefit of disenfranchised rural communities are addressed. The lessons learned are instrumental to the successful implementation of future development investments.'

Professor Judi W. Wakhungu, African Centre for Technology Studies, Nairobi, Kenya

Contents

List of illustrations

Figures

Tables

List of abbreviations

ACDI/VOCA Agricultural Cooperative Development International/ Volunteers in Overseas Cooperative Assistance
ACTS African Centre for Technology Studies
AECF Agriculture Enterprise Challenge Fund (DFID)
AFRISA Africa Institute for Strategic Animal Resource Services and Development
AGRA Alliance for a Green Revolution in Africa
ASARECA Association for Strengthening Agricultural Research in Eastern and Central Africa
AVRDC Asian Vegetable Research and Development Centre (now known as the World Vegetable Centre)
BBSRC Biotechnology and Biological Sciences Research Council
BCA Bio Control Agent
BecA Biosciences east and central Africa
BMGF Bill and Melinda Gates Foundation
BVM Bachelor of Veterinary Medicine
CAADP Comprehensive Africa Agriculture Development Programme
CABE Centre for Africa Bio-Entrepreneurship
CABI Centre for Agricultural Bioscience International
CAII Centre for Agricultural Industry Ltd
CBAF Community-Based Armyworm Forecasting
CBO Community Based Organisation
CDF Community Development Fund (Kenya)
CEPA Cambridge Economic Policy Associates
CGIAR Consultative Group for International Agricultural Research
CIAT International Centre for Tropical Agriculture
CIMMYT International Maize and Wheat Improvement Centre
CORAF Counseil Oust et Centre Africain pour la Developpment Agricoles (now known as West and Central African Council for Agricultural Research and Development (WECARD)
CPP Crop Protection Programme (under the DFID RNRRS)
CRD Central Research Department (DFID)
CRT Central Research Team (DFID)

CSR Corporate Social Responsibility
CTVM Centre for Tropical Veterinary Medicine
DADP District Agricultural Development Plan
DDS Diocesan Development Services
DFID UK Department for International Development
DHSS Department of Health and Social Security
DLCO Desert Locust Control Organisation
DTI Department of Trade and Industry
DVO District Veterinary Officer
EPA Environmental Protection Agency
EPSRC Engineering and Physical Sciences Research Council
FAO Food and Agriculture Organisation
FFAI Food for All International
FFEPP Fish Farming Enterprise & Productivity Programme (Kenya)
FFS Farmer Field School
FIPS Farm Input Promotions Ltd.
FVM Faculty of Veterinary Medicine
FYS Farmer-Yield-Share
GDP Gross Domestic Product
GEF Global Environment Facility
GLOBALGAP Global Partnership for Good Agricultural Practice
GoK Government of Kenya
GoT Government of Tanzania
GPS Global Positioning System
GSI Good Seed Initiative
HAT Human Acquired Trypanosomiasis
ICIPE International Centre for Insect Physiology and Ecology
ICRISAT International Centre for Research in the Semi-Arid Tropics
IFAD International Fund for Agricultural Development
IFPRI International Food Policy Research Institute
IITA International Institute of Tropical Agriculture
ILRI International Livestock Research Institute
IPM Integrated Pest Management
IRRI International Rice Research Institute
I4D Innovation for Development
JICA Japanese International Co-operation Agency
KARI Kenya Agricultural Research Institute
KEPHIS Kenyan Plant Health Inspection Service
KPL Kilombero Plantations Limited
LIC Low Income Country
MoA Ministry of Agriculture, Kenya
MoAFS Ministry of Agricultural and Food Security, Tanzania
MoARD Ministry of Agriculture and Rural Development, Ethiopia
MAFSC Ministry of Agriculture and Food Security, Tanzania
M&E Monitoring and Evaluation

MDGs Millennium Development Goals
MITI Ministry of International Trade and Industry (Japan)
MINTRACS Makerere in-Training Community Service
MRC Medical Research Council
MSHR Missionary Sisters of the Holy Rosary
MSVD Maize Streak Virus Disease
MTR Mid-Term Review
NAADS National Agricultural Advisory Services (Uganda)
NACRRI Uganda National Crops Resources Research Institute
NARS National Agricultural Research Systems
NARO National Agricultural Research Organization
NCCI National Commission on Cohesion and Integration
NEPAD New Partnership for African Development
NERICA New Rice for Africa
NRCRI Nigerian National Root Crops Research Institute
NRI Natural Resources Institute, UK
OAF One Acre Fund
OAS One-Stop-Aqua-Shop
ODA Overseas Development Administration
NGO Non-Governmental Organisation
NRCRI National Root Crops Research Institute (Nigeria)
OST Office of Science and Technology
PLOSNTD Public Library of Science Neglected Tropical Diseases
PPARC Particle Physics and Astronomy Research Council
PPD Plant Protection Division, Kenya
PCPB Pest Control Products Board, Kenya
PEC Protein Engineering Club
PPP Public Private Partnership
R&D Research and Development
RAP Restricted Application Protocol
RIU Research into Use
RSA Republic of South Africa
RNRRS Renewable Natural Resources Research Strategy
SAACO Safe and Affordable Armyworm Control Tools
SADC Southern Africa Development Co-operation
SERC Science and Engineering Research Council
S/T Science and Technology
SFSA Syngenta Foundation for Sustainable Agriculture
SME Small and Medium-Sized Enterprise
SMS Short Message Service
SOS Stamp Out Sleeping Sickness
SRSA Strategy for Research on Sustainable Agriculture
SSA Sub-Saharan Africa
SYCI Shujaaz Youth Communications Initiative
TOT Transfer of Technology

TR Technical Review
UNCED United Nations Conference on Environment and Development
UNDP United Nations Development Programme
UNESCO United Nations Educational, Scientific and Cultural Organisation
UPOV International Union for the Protection of New Varieties of Plants
USAID United States Agency for International Development
VBA Village Based Adviser
WARDA West Africa Rice Development Association
WECARD West and Central African Council for Agricultural Research and Development
WTS Well Told Story Ltd.

Preface and acknowledgements

A question that continues to be debated in relevant policy circles is that of the effectiveness of scientific research in improving rural poverty in low income countries. Since the advent of the Green Revolution 50 years ago or so, considerable sums of public resources have been spent on research. Recent estimates suggest that a sum of more than $7 billion has been spent in international research centres alone between 1960 and 2001 (Conway & Waage, 2010). And certainly some of this effort has translated into many useful technologies that directly and indirectly improve the lot of the poor farmer. Indeed evidence suggests that crop yields in developing countries would have been up to 23.5 per cent lower, with food prices some 35 per cent to 66 per cent higher, had it not been for the technological impacts of the Green Revolution (Juma, 2011). And yet for the poorest farmers, especially in African regions, overall impact has remained poor. Of course part of the problem is one of contextual conditions, such as poor soils, insufficient economic infrastructure, poor government policies and water stress. But nevertheless, the uneven impact (by region and income class) suggests that there are also problems of effective translation of scientific knowledge into productive practice.

This monograph is about an attempt to throw more light on this question by a bilateral aid agency, the UK Department for International Development (DFID); it summarises one part of a large development research programme funded by DFID's Central Research Department (CRD).[1] The Research into Use (RIU) Programme originally budgeted at £37.5 million, began in July 2006 and lasted until December 2012.[2] The Best Bet component of the programme on which this monograph is based started in mid-2009. It was budgeted at £5 million and was specifically designed as an experiment focused on leveraging private sector capacity in putting previously funded (1995–2005) DFID research under its Renewable Natural Resources Research Strategy (RNRRS) into developmental use. The RIU has broken with tradition in at least two respects. Firstly, it was established at least partly as a learning activity; secondly, and in consequence, it has had a research-action character. In other words, the learning outputs could not be exactly anticipated since they depended on results that were unknown at inception. What were needed were results that would indicate to the CRD how it might develop more appropriate research policies for technology development aid to low income countries targeted at rural populations. In turn, this aim had been prompted by concern that much of its previously funded research was not actually being used; or at least it was hard to demonstrate use unambiguously.

The monograph's authors are those that were most concerned latterly with the Best Bet part of the RIU. Professor Ian Maudlin and Dr Andy Frost were respectively Director and Deputy Director of the RIU based at the University of Edinburgh. Professor Norman Clark (Open University) acted as the programme's economic adviser with special responsibility for the Best Bets. He was assisted by Dr Andrew Ward (at that time located at the CGIAR Consortium Office in Montpelier in France), who collated and analysed on-going project details. However, mention should also be made of the wider RIU staff who were closely involved, especially since 2009. Professor Andy Hall acted as leader of the Central Research Team (CRT) with responsibility for RIU learning as a whole. The CRT team and its research fellows produced some 27 Working Papers that may be accessed at the RIU website (see also Appendix 1);[3] many of these papers acted as sources for this monograph and some are now appearing as journal articles in their own right. Duncan and Keith Sones ran the RIU communications activities, including running and managing its entire programme and related meetings. Considerable assistance has been given by Best Bet project staff and many of the scientists involved in the original RNRRS research.

We are grateful to the wide range of personnel from many institutions who gave interviews and related help in putting this monograph together. In particular mention should be made of colleagues at the African Centre for Technology Studies (ACTS), Nairobi, the Development Policy and Practice Group in the Faculty of Technology at the Open University (OU) and the ESRC-supported INNOGEN Research Centre (University of Edinburgh and the OU). In addition our thanks are due personally to the following: I. Barker (Syngenta Foundation), N. van Beek (Kenya Biologics Ltd), M. Bell (University of Sussex), R. Burnet (WTS Ltd), D. Campbell (Mediae Ltd), J. Chataway (Rand and OU), J. Cordingley (Crop Protection Laboratory Services Ltd), D. Coyne (IITA), N. Daniels (One Acre Fund), R. Day (CABI), M. Farroe (USAID), P. Gildemacher (KIT), D. Grzywacz (NRI, Greenwich University), D. Harris (ICRISAT), S. Kariuki (ICIPE), M. Kett (LCDDC), J. D. Kabasa (FVM, Makerere University), A. Kingiri (ACTS), A. Kyalo (Kenya Ministry of Agriculture), L. Labuschagne (Real Impact for Sustainable Growth), N. MacNamara (MSHR), S. Morse (University of Surrey), A. Mufuruki (Infotech Investment Group), J. Muema (LCDDC), S. Murunga (ACTS), B. Muok (ACTS), R. Musebe (CABI), W. Mushobozi (Eco Agri Consult Ltd), S. Nderitu (ACTS), L. Nickoll (ESPA), F. Nkirote (ACTS), H. Odame (CABE), P. Oketa (African Agricultural Capital), G. Osure (Syngenta Foundation), S. Otieno (Farm Africa), N. Phiri (CABI),

D. Priest (FIPS-Africa Ltd), A. Radcliffe (World Agroforestry Centre), R. Rajalahti (World Bank), P. Robbins (OU), P. Seward (FIPS-Africa Ltd), P. Toye (ILRI), M.Wahome (Deacons (K) Ltd), H. Wainwright (Real IPM Ltd), J. Wakhungu (ACTS), A. Watt (Syngenta AG), C. Waiswa (FVM, Makerere University and High Heights Ltd), S. Welburn (School of Biomedical Sciences, University of Edinburgh), D. Wield (INNOGEN), K. Wilson (Lancaster University).

Finally we would like to stress that this monograph has been written to be accessible to the wide range of students and development practitioners who may not be directly concerned with the narrower issues of technology development in poor countries. Those interested in the broader context of poverty related issues in Africa are referred to Conway (2012) and Juma (2011). Conway's book provides a comprehensive, up-to-date account of the need for international food security, the potential for achieving it and the constraints confronting it. Juma focuses directly on African agriculture emphasising the emergence of new types of political leadership, the creation of regional markets and emergent entrepreneurial capacities.

Conversely, this monograph focuses more narrowly on technology development possibilities for the small farmer. The projects dealt with cover a wide range of interventions from crop protection to animal health; and so we have tried to compromise by presenting enough summary detail to inform the non-specialist reader while at the same time not getting him/her bogged down in what have often turned out to be complex stories. Hence, the substantive accounts (Chapters 4–6) of the nine Best Bet projects (which are still on-going) have been written descriptively to provide sufficient information for a broad readership who may not wish to go into finer project details.

Materials used have been taken from internal project documents supplemented by interviews with relevant personnel. Direct sourcing of data has been largely omitted. However, those wishing further details are encouraged to access the website material outlined in Appendix 1. This provides a direct link to the 27 papers in the RIU discussion series put together by Professor Hall's Central Research Team, plus other multi-media links to the projects themselves. It also includes links to specific M&E material commissioned by RIU management.

This document is an output from the Research into Use Programme funded by the UK Department for International Development (DFID) for the benefit of developing countries. The views expressed are not necessarily those of DFID.

Notes

1 This has now been renamed as the DFID Research and Evidence Division (RED).
2 Comprehensive details of RIU may be accessed through www.researchintouse.com.
3 See http://researchintouse.com/learning/learning40discussionpapers.html.

1 Technology development for agriculture

Broad context

Introduction

This monograph explores aspects of technology development aid to low income countries (LICs). It provides an analysis of parts of a wider technical assistance programme designed to improve the impact of scientific research on agricultural production with a special focus on the rural poor. The UK Department of International Development's *Research into Use* (RIU) programme was established both to improve the use of its previously funded research in LIC agriculture and to better understand the factors that are relevant in this respect. The monograph concentrates on what came to be called the Best Bets, a small set of projects within the wider programme designed to leverage private sector input into technology development, thereby encouraging (hopefully) greater sustainability than would be obtained with more conventional types of aid. These projects have been conducted in Africa over the period 2009–2012.[1] Some are now completed; others are nearing completion. As outlined in the Preface, the following chapters cover science policy analysis, the projects themselves, their selection, how they fit within the wider RIU programme and most importantly what their implementation appears to tell us about this type of technology development aid.

This chapter gives a short overview of the operational context. Its main argument is that while the application of scientific and other knowledge to poverty related agriculture has been on the agenda for at least 50 years, impact on African agriculture has not been impressive.[2] This was an important factor influencing the inception of the RIU and indeed the RIU has represented an institutional break with traditional approaches to the use of knowledge in development, certainly with respect to agriculture. Arguably, the ruling paradigm in this context is still the *Green Revolution* model whereby centrally conducted science, in specialised organisations, produces generic knowledge that is then transferred to other bodies until ultimately it receives expression in the fields of the poor farmer. What the RIU has essentially been about is to suggest that there are other ways of achieving technology development and that these should now be explored. This does not mean the downgrading of science's role. Nor does it deny the usefulness of existing structures,[3] but it does imply that the institutional context for agricultural science may need some revision. The RIU addressed this issue in the context of aid to the natural resources sector.

The paradox of Africa's agriculture has been much written about in recent years. Despite the abundance of natural resources, significant investments in R&D and continuous remedial injunctions in policy circles, its relative stagnation over recent decades has been unrelenting. Indeed it amounts in CAADP's view to a continental crisis affecting particularly the rural poor. At its inception in 2003 the NEPAD website put it as follows:

> Farmers' yields have essentially stagnated for decades. Although total output has been rising steadily – often by simply extending the land area under cultivation – this growth has barely kept pace with Africa's increasing population. Food production in particular has lagged, so that the number of chronically undernourished people increased from 173 million in 1990–92 to 200 million in 1997–99, the latest years for which accurate figures are available. Of that total, 194 million were in sub-Saharan Africa. This growth in hunger has come despite high levels of food imports – costing $18.7bn in 2000 alone.[4]

It would be tedious to detail yet again all the relevant causes and factors involved.[5] They range from insufficient investment, declining farm sizes, uncertain rainfall and poor access to markets, to weak social infrastructures, poor national government policies, corruption, the impact of structural adjustment measures and a hesitant response on the part of donor agencies. And yet as Juma (2011) points out, the development of new technologies and related innovation holds out promise for significant improvement. The question is, what reforms in international science policy are needed to realise this promise?

Sixty years ago the scientific approach was very positive. The end of the Second World War (WW2) heralded a period of international optimism that combined the recognition of economic inequality with a determination that international action could reduce it substantially. And while the organisations created at the time (such as the United Nations and the big financial bodies like the International Monetary Fund and the World Bank) probably had peace and stability as their primary focus,[6] an important secondary element was the harnessing of science and technology (S/T) to economic growth and development. The early debates on this centred on the disjunction between modernisation and autarchic agendas. Initially the former held sway but as we moved into the 1970s the tone became much more radical (in the old political sense),[7] technology being seen as a tool of exploitative economic relations on the part of corporations having their roots in the industrialised countries of Europe, North America and East Asia.

The influence of modernisation policies was significant however, since they laid the basis for institutional investments in many countries (including former colonies). Since technology was a key source of economic growth (this had been shown unambiguously by economists in the 1950s and early 1960s;[8] it was subsequently given greater formal standing with the development of 'new growth theory' in the 1980s) it should clearly be harnessed in the service of development. In policy terms, however (and guided by international bodies like UNESCO), the goal was seen to be one of creating universities, research institutes and other S/T bodies in the mould of those already extant in the industrialised countries. For agriculture this investment primarily took the form of what became the Consultative Group for International Agricultural Research system (CGIAR), through which scientific resources were concentrated in centres of excellence that had led to what later became the *Green Revolution*. This chapter will take this as its starting point before looking more broadly at how this has played out in the context of technology development in agriculture.

CGIAR or CG System

As mentioned above, the early approach to science and agriculture was conditioned by the aftermath of WW2. Political rivalry between capitalism led by the USA and Soviet socialism had begun, and underdevelopment showed a likelihood of becoming one of its ideological battlegrounds. At the time, American foreign policy centred on the control of communism, which was believed to thrive under conditions of underdevelopment in general, and food insecurity in particular. The United States had built up extensive expertise in plant breeding, which was credited with having led to significant agricultural improvements in the southern United States. Building upon these successes, and concerned with promoting American strategic interests abroad, two private philanthropic organisations – the Rockefeller and Ford Foundations – began to explore means of extending American agricultural expertise to 'underdeveloped' regions.[9]

The context of these developments included American Cold War era concerns about its national security and the perceived threats from poorer countries where rising populations were placing a strain on food security and making them vulnerable to communist takeover. The key actors involved in articulating the link between American national security and development assistance in poorer countries formed a social network through professional and personal ties, which stretched to the American political establishment. At the centre of this network was the Rockefeller Foundation, instrumental in couching population, food productivity and conservation issues in a language that resonated with United States' political priorities.[10] Smith (2009) puts it as follows:

> (p)artly because of the faith placed in science and partly because of the idea that the application of new technologies was a universalising process, great emphasis was placed on using existing expertise to solve isolable technical problems. Such a perspective involved removing the research process from its context not once but twice. Firstly, a particular North American perspective on how to undertake agricultural research was proposed as a model for the approach, and secondly the problems themselves were to be solved out of context; the problem of yield was taken to stand for the problem of production and the problem of production was taken to stand for the 'world food problem'. Yields of rice and other crops had doubled in the past fifty years in the United States. As the causes of these increases were at the time well known, generating similar increases in other countries and contexts would simply be a matter of direct technical intervention.

He goes on to point out that the focus of what then became known as the *Green Revolution* was one of applying concentrated scientific resources to develop new varieties of rice, wheat and other cereals that would respond well to other inputs (particularly water and artificial fertiliser) and thereby dramatically improve production possibilities in LICs. These crop varieties initially came from the International Rice Research Institute (IRRI) and the International Maize and Wheat Improvement Centre (CIMMYT). The immediate improvements were considerable. For example, Conway and Waage (2010: 67) show that between 1961 and 1985 cereal production in developing countries more than doubled. But while benefits have since been apparent in much of Asia, the Middle East and Latin America, large parts of South Asia and virtually all of Sub-Saharan Africa have missed out. The World Bank's comprehensive *Development Report* (2008: 4) concluded that 'the large decline in the number of rural poor (from 1,036 million in 1993 to 883 million in 2003) has been confined to East Asia and the Pacific (while) in Sub-Saharan Africa, the number of rural poor has continued to rise and will likely exceed the number of urban poor until 2040'. Indeed Conway and Waage maintain that on present trends there is little likelihood of Sub-Saharan Africa meeting the millennium development goal (MDG) of halving the proportion of hungry by 2015; Ghana is the only exception.[11]

The CG System itself was established in 1971 as an 'informal association of governments, international organisations and private institutions co-sponsored by the World Bank, the Food and Agriculture Organisation (FAO) and the United Nations Development Programme (UNDP)'. Its objective was 'through international agricultural research and related activities, to contribute to improved sustainable food production in LDCs in such a way that the nutritional level and general economic well-being of their low-income peoples are improved'.[12] The immediate target population was therefore the rural poor – i.e. low-income, semi-subsistence farmers – and the mandate was one of overcoming technological constraints on increasing output of agricultural commodities. The CG System started with a nucleus of four research institutions, two of which (CIMMYT and IRRI) as we have mentioned had become famous for their contributions to the Green Revolution – the development of high-yielding varieties of wheat and rice which had begun to transform economic possibilities in many parts of the Third World. The other two (the International Centre for Tropical Agriculture [CIAT] and the International Institute of Tropical Agriculture [IITA]) were established later on in the 1960s.

Over the past 35 years the CG System has grown to some fifteen centres and 64 governmental and nongovernmental members (by 2011). In terms of research focus, its centres are mandated to conduct 'generic' or 'strategic' research related usually to a specific commodity or agro-ecological zone although there are a small number, such as the International Food Policy Research Institute (IFPRI) that are more broadly institutional/policy in orientation. There is often a concentration on biological research and training associated with a farming systems approach to agricultural development; and within this broad area many centres are paying a lot of attention to genetic resource conservation and biotechnology research. As a consequence of its most recent reform the new CGIAR research programmes are now expected to be responsible for bringing about development and environmental improvements. Traditionally, however, they continue to be seen as centres of research excellence producing generic technologies which can be applied in specific geographical contexts. The job of actually applying the technologies and transferring them to farmers is normally left to the National Agricultural Research Systems (NARS), though since some NARS are still weak, CG institutions sometimes get involved more 'downstream' – i.e. to strengthen the NARS institutionally and help them build linkages with other components of the global agricultural system. Nevertheless, the overall strategy of the CG System has tended to favour 'upstream' activity with fairly clear injunctions to concentrate on publishable research.

Changing agricultural research context

Given prevailing conditions at the time, it is clear that the CG System represented an important set of aid interventions, since despite international help for national agricultural research bodies, many developing country governments did not support local research sufficiently. In compensation, therefore, multilateral and bilateral aid, along with support from private foundations, was channelled to international centres of excellence that would undertake strategic research for developing countries without becoming enmeshed in the administrative and political arenas of client countries. As outlined above, what emerged was a two-tiered system. The CG centres developed production technologies and varieties for mandated crops (and geographical regions) that were subsequently passed on to the NARS for applied, contextual research and final transfer to the farmers through extension agencies. This essentially is the 'transfer of technology' (TOT) model, which became the 'engine' of the *Green Revolution* and which has really dominated policy thinking until very recently. Its chief characteristics were a belief in the existence of scale economies in the R&D process, a faith in the scientific method as the main source of improved technological practices for the poorest of the poor, relatively little attention paid to other knowledge sources, or the tacit knowledge and local preferences of the farmers themselves.[13]

However, the issue of rural poverty in poor countries persists, even, in places, getting worse, with growing recognition that there are large parts of the world's population that are still living in dire poverty and under poor and worsening environmental conditions. It was mainly to address this issue that the Millennium Development Goals (MDGs) were adopted by the United Nations General Assembly in 2000. These were articulated by a number of Task Forces into steps that need to be taken if these goals (and related targets) are to be achieved over the coming years.[14] But if scientific knowledge is to contribute to this aim then clearly the TOT model will need to be re-visited, since despite much scientific progress having been made, the actual impact on poverty alleviation, especially in the African context, remains marginal. It seems therefore that while disinterested research in well-funded institutions can play an important role, more is required. And there is some indication that this is being recognised. For example, the role of science in improving agricultural development was addressed in 2004 by the then President of the UK Royal Society (Lord May) as follows:

> Take for instance the goal of eradicating extreme poverty and hunger. There are economic, political and cultural social science elements to achieving this aim. But the physical and biological sciences and technologies will also be crucial, for instance in improving agricultural methods and technologies, food preparation and distribution techniques and systems, and building up basic education about nutritional needs … Quite simply, without significant scientific infrastructure and expertise within the poorest countries, it will be difficult if not impossible for them to help themselves in finding solutions. This will make them ever more reliant on aid and assistance from more scientifically developed nations. In other words, without the capacity embedded in the educational system at school, university and in research laboratories, and without the institutions such as academies to provide the support structure for good scientists to work in the countries concerned, there is little chance that science will be able to be harnessed to address these countries' needs, or even that these countries will be able to absorb and use the science generated elsewhere.[15]

Implicitly, May is arguing that science as such is only part of the story. What are also needed are new types of institutional practices that enable knowledge in its wider form to have a more direct impact. But what are these and how should they be introduced? A useful starting point is to outline how contexts have changed since the Green Revolution. The following are those most often provided in the literature.[16]

The macroeconomic context

There have been radical changes regarding the role of the state in socio-economic governance. In simple terms, the approach has been as far as possible to withdraw from direct state control of the economy, shifting emphasis away from state implementation to that of providing an appropriate (macro) policy environment. Central to this is the move to allow the market to provide services and to use competition to generate efficiencies that the public sector arguably cannot achieve. The developing world first felt the consequences of the new ideology in the structural adjustment packages implemented by international financial institutions in the 1980s. The 'adjustments' referred to macroeconomic and trade policy reforms (such as exchange rate reform for example) but they also had to do with changes in the structure of the economy, mainly the extent to which the state provided public services and controlled key economic sectors. The approach was an attempt to reduce large and apparently unproductive public sector bureaucracies, to break up state monopolies and to open up markets to competition, both nationally and internationally. As a result agricultural research systems began to face new challenges, particularly in agricultural research and extension funding which in Africa has declined considerably as a consequence (Pardey *et al.*, 1997). Now, as we come back to later in the text, formal extension systems simply cannot deal with what is required of them and a variety of other types of *ad hoc* advisory services have emerged to fill the gap.

Private sector

A key change that is making this all the more pertinent is the emergence of private sector research. This was partly a result of improved intellectual property protection regimes and the technical advances associated with biotechnology. But also significant are the opportunities that economic and trade liberalisation and globalisation are now presenting for private investments in agro-industries such as seed production (Pray, 2002; Muraguri, 2010). The net result is that public agricultural research systems have to consider more modified roles. But these may often conflict with traditional internally driven policies and beliefs of the research sector, particularly where these remain focused on production and productivity and continue to reflect the food security concerns of an earlier period

(Roseboom and Ruttan, 1998). Increasingly, the difficulty concerns the management of more complex situations, with policy makers having to choose between serving, for example, the commercial needs of the agricultural sector while simultaneously serving the interests of society at large. In this way agricultural policy is much more complex than it used to be (Hall & Yoganand, 2004; Hall, 2009).

Poverty impacts and donor responses

While in earlier periods it is clear that big productivity gains were achieved, particularly with respect to maize, wheat and rice, as time has gone by the evidence on real poverty impact has become much patchier. In particular, Green Revolution technology tended to require high levels of ancillary inputs such as water and fertiliser, and supportive institutional structures dealing with extension, credit and marketing. Richer farmers had access to these but poor farmers relied heavily on state support such as subsidies for key inputs like fertilisers. Such support was forthcoming in the 1970s but as global financial crises impinged in the 1980s it began to vanish. In this context it is important to note that while earlier criticisms of agricultural research performance concerned the appropriateness of the new varieties and technology in terms of their suitability for poor farmers, this criticism left the agricultural research system largely blameless. That is, application questions were assumed to be 'exogenous' to research.

However during the 1980s increasing criticism emerged of this (institutional) model of agricultural research. Biggs (1990) in particular drew attention to the key institutional dimension of the problem and specifically the hierarchies inherent in agricultural research systems fashioned on the 'linear' model of innovation. Nowadays many donors are less willing to fund research that does not have operational links with application to the poorest sections of rural society where food security is still a major issue. They are rather more inclined to support projects that show clear impact on rural livelihoods and conversely less keen to support agricultural research for its own sake. In other words, they are less willing to tolerate the 'assumptions' column of the log frame matrix. A good example is that of a USAID project to upgrade sorghum and millet research in the Africa SADC region. Hall *et al.*, (2004) show that the donor in this case was unwilling to accept only the development of new plant varieties but demanded also to be shown quantitative evidence that these had actually been adopted by farming communities. In this case, the scientists themselves enlisted the support of local NGOs to ensure success in this wider sense.[17]

Environmental issues

Over the past 50 years the issue of environmental degradation has become much more significant. In particular intensive agricultural development often relies on chemical inputs and heavy consumption of water. A specific example is the salinity effects of large-scale use of ground water. A more general issue is the falling of the water table in areas where recharge capacity is hampered by vegetative depletion combined with new types of land use. Many of these issues have focused recently on questions about how to respond to the possible impact of climate change.[17] Another area that has caused popular concern is the use of transgenic technology in plant breeding and livestock production. The watershed here was probably the Biodiversity Convention at the Rio UNCED 'Earth Summit' in 1992 and the subsequent ratification of its Biosafety Protocol by most African countries between 2002 and 2004. But the impact has had far reaching consequences, particularly in export sectors where access to markets can often be restricted.

There are also implications for the extent to which poor countries should accept or allow human consumption of genetically modified food. At present, many African governments have now formulated regulations on the handling, development, transfer and use of modern biotechnology, particularly genetically modified products. They are doing this to respond to international policy and law. For example, with financial support from the Global Environment Facility (GEF) more than 15 African governments have developed guidelines and laws on biosafety to ensure the safe development and application of biotechnology. These are important issues and indeed most African countries are now putting in place biosafety policies with clear implications for the pattern and effectiveness of agricultural research. However, it is not clear to what extent local and international scientists are participating in informing such policy processes.[19]

Biotechnology

The knowledge intensity of all economic production has increased due to technological developments. For agriculture, however, biotechnology is of central importance, although it is still the subject of controversy.[20] For example, on the one hand, biotechnology promises the capacity to radically improve rates of growth of food production and of other primary commodities such as cash crops for export. It can also help reduce environmental damage through curtailing the use of chemical pesticides and herbicides, and help deal with problems of growth stress. On the other, there are dangers that new synthetic substitutes derived from biotechnology can drive traditional export products out of the market. For example, companies based in the north can produce products like pyrethrum and artificial sweeteners without any recourse at all to traditional products and the chances are that this capacity will grow considerably over the coming decades.

Similarly with the use of modern techniques of tissue culture. Wambugu and Kiome (2001) show how tissue culture has been used to promote the production of disease-free bananas in East Africa. The potential benefits for many subsistence farmers are likely to be considerable. On a more industrial scale, as noted above, tissue culture is now being used to promote the production of export-led high value horticulture crops (Bolo, 2012), although international markets (particularly the EU) are calling into question environmental and social methods of production. In addition, concerns have been expressed in recent years regarding the way international seed corporations have begun to dominate agricultural production in many developing countries, for example through using genetically engineered seeds in a proprietary fashion. All this is not to say that the agricultural 'knowledge economy' is confined to advances in the new 'genomics'. On the contrary, there are many other areas where traditional agricultural science is equally important. The point is rather that agricultural science systems need to build the capacity to select what are the most effective mechanisms to invest in so that the impact of their R&D programmes is at its greatest.

Stakeholder groups

Traditionally the job of passing on the results of agricultural research was given to state-funded extension services. Not only have these suffered through structural adjustment measures but there are also increasing questions raised about the extension systems themselves as an operant organisational mechanism.[21] There is also evidence of increased need to engage in partnerships in order to deliver new technologies successfully to client groups. These partners include private sector organisations but they also involve NGOs and community-based organisations (CBOs) that are able to bring skills and knowledge to bear simply due to the close relationships they have established with specific communities. Hall *et al.* (2004) and Hall (2006; 2009) summarise some of the increasing number of studies that show how the ability of new technologies to impinge directly on poor farming communities has been considerably enhanced through the use of well-selected NGOs.

Farmer knowledge and 'participation'

A related issue concerns how to involve farmer knowledge in R&D activity. Starting in the mid-1980s there has been a growing acceptance of the necessity to involve the 'client' not only in 'needs definition' but also in how local conditions are likely to affect agricultural interventions. For example, a study of hybrid maize in the Siaya region of West Kenya shows that despite vigorous (and costly) promotion of hybrid maize on the part of scientific and government authorities over a twenty-year period, 'local' varieties continue to remain widely grown while hybrid varieties are hardly grown at all.[22] The reasons for this are fundamentally ecological. Local 'landraces' fulfil needs associated with pest and water tolerance, costs of expensive inputs such as fertilisers, product acceptability on the part of consumers, productivity of 'saved seed' and credit scarcity. Conversely, modern hybrids need reliable rainfall, expensive inputs, rigorous planting schedules and timely labour availability – in fact the entire supportive institutional framework that exists on the typical experimental station. It is clear that formal research systems would benefit from a greater awareness of such problems but they still have difficulty engaging with the issue.

In recent years agricultural research systems have begun to try and accommodate (with varying levels of enthusiasm and success) this new agenda in both research practice and research focus (see for example Hall and Nahdy, 1999). However, all too often much of the advocacy for this change in approach has focused on participatory methods rather than underlying institutional issues.[23] Biggs and Smith (1998) argue that this 'methods bias' masks the fact that the most successful participatory methods have arisen in specific institutional and political circumstances and have often evolved to deal with a specific problem area in that context. In addition they have often been characterised by a significant degree of institutional innovation. Hall and Nahdy (1999) argue that agricultural scientists all too frequently find themselves struggling to apply participatory approaches in an institutional and professional context that implicitly denies such patterns of client interaction.

The new agriculture (and challenges of creating dynamic innovation capacity)

Part of the problem that agricultural research faces is the fact that the 'one-size fits all' model of an agricultural research system is simply not suited to the emerging reality of the developing country agricultural sector (Anderson, 2008). While production, sale and consumption of major food crops remains important, a number of niche sectors are emerging with impressive rates of growth and this is coupled with fundamental changes in the nature of the sector as whole. As Hall *et al.* (2004) have pointed out in some detail, these include the growing importance of the livestock industry, particularly aquaculture; the diversification into horticulture and cut flowers, particularly for export; the use of roots, tubers and coarse grains for food and industrial use (animal feeds, bio-fuel, starch); increased consumption of processed foods and the growth of a processing industry; the increasing role of the private sector; corporatisation of craft-based industries such as herbal medicines; the exposure of producers and firms to competition, changing international trade rules and regulations such as sanitary and phytosanitary standards; the knowledge-intensive nature of these niche sectors; and the importance of innovation as a source of competitive advantage in rapidly evolving market and technological conditions. Factors driving these changes include globalisation of markets, rapid urbanisation, changing food preferences and the industrialisation of the food chain

The evolution of such sectors is not necessarily going to benefit the poor in the traditional way of providing new opportunities to the poor as farmers – although it does not preclude that. Instead, it suggests that rural non-farm employment opportunities will become more important. Take for example cut flowers in Kenya. Not only did it achieve a five-fold increase in production and sales between 1997 and 2006, but it is highly labour intensive, employing some 70,000 mainly women workers. It is also the third best foreign exchange earner after tea and tourism (Bolo, 2012).[24] Another example is the aquaculture sector, which has expanded rapidly in many Asian countries showing impressive rates of growth (Conway and Waage, 2010). A recent conference organised by the FAO (2011) concluded that 'production of fish from aquaculture has grown substantially in the past decade, reaching 52.5 million tonnes in 2008, compared with 32.4 million tonnes in 2000'. Aquaculture continues to be the fastest-growing animal food producing sector and currently accounts for nearly half (45.6 per cent) of the world's food fish consumption, compared with 33.8 per cent in 2000. With stagnating global capture, fishery production and an increasing population, aquaculture is perceived as having the greatest potential to produce more fish in the future to meet the growing demand for safe and quality aquatic food.[25]

A new agenda?

In short, modern literature shows that the context for LIC agricultural research has changed considerably from the days of the *Green Revolution*, and with it the demands on relevant institutions. It is this changed context that has prompted a fresh look at science policy analysis for agriculture. Arguably, however, technology development still depends heavily on traditional structures and habits. The centre of gravity is still the large research institution producing knowledge that has difficulty reaching the ultimate user, the poor farmer in unsupported rural regions of Africa. This is not to say that attempts are not being made to improve access but it is slowly becoming recognised that much in the agricultural extension world is not fit for purpose. As Anderson (2008: 27) points out:

> Many administrative and design failures have proved problematic in public extension effort in the past, most notably those associated with: the scale and complexity of extension operations; the dependence of success in extension on the broader policy environment; the problems that stem from the less than ideal interaction of extension with the knowledge generation system; the difficulties inherent in tracing extension impact; the profound problems of accountability; the oftentimes weak political commitment and support for public

> extension; the frequent encumbrance with public duties in addition to those related to knowledge transfer; and the severe difficulties of fiscal unsustainability faced in many countries. Wide-ranging reforms have been experimented with and new approaches introduced in many parts of the world to do a needed job better, (but) there is clearly a need for more empirical research to find out 'what works where and why'.

And there is certainly considerable debate. An important part of this lies in the structure of the sector itself. There are some like Collier (2008) who believe that there is little future for peasant smallholder agriculture as currently practised. In his view, the answer lies exclusively in the realm of large scale (mainly commercial) farms. Anything else is dismissed as 'romantic populism' and is bound to fail. Others like Conway (2012), Hazell *et al.* (2010) and Wiggins *et al.* (2010)[26] believe that there is still a bright future for smallholder agriculture, while Letty *et al.* (2012) agree but are uncertain about how to proceed in terms of innovation policy. Increasingly, however, it is recognised there is a need for fresh thinking. For example, bodies like the Syngenta Foundation for Sustainable Agriculture (SFSA) have promoted a range of experimental schemes in recent years designed to minimise risks and expand technical opportunities.[27] Their view is that a middle road should be found between large and small scale interventions, one that takes advantage of economies of scale where possible but at the same time providing scope for the smallholder to upgrade technologically and economically. Recent examples in east Africa are as follows.

SFSA funding is provided for the *One Acre Fund (OAF)*, a US-based group established and run by an ex-Harvard Business School graduate.[28] It operates through loan packages of around KSh 4,000 ($40) provided to small farmers. These include seed, fertiliser, agronomic advice and marketing. Farmers are clustered in 'farmer groups' and the OAF supports the farmer groups through a network of advisers. The OAF has been working in Bungoma since 2006 and claims now to have reached some 24,000 farmers. It also goes well beyond conventional farming, supporting also bed-net provision and tree planting. Since 2009, SFSA has developed the *Kilimo Salama* insurance scheme to compensate for bad weather fluctuations. It can be bought through agrovets (also called agro dealers) and is monitored through a patchwork of small scale weather stations which sanction payout based on the actual amount and distribution of rainfall over the crop season. It uses IT technology and seems an efficient mechanism for insuring the small farmer against weather fluctuations. By 2011 the scheme claimed to have reached 11,000 farmers and there are now plans to expand to cover additional risks.[29]

SFSA also supports partnerships with the private sector on seed supply, vegetable nurseries and marketing, generally trying to assist the extension system wherever possible. The parent company Syngenta AG also believes that there is a role for partnering the small farmer with larger operations. A good example of this is the large (5,000 ha) rice farm in Tanzania (KPL). In addition to its normal operations KPL provide contract farming facilities to local smallholders. These cover access to credit, training, rice milling, clean input provision and price guarantee. They have started with 150 farmers but intend to expand this number substantially. There is also provision for marketing under the KPL brand. Syngenta AG believes that CSR activity of this type will pay off in the long run.[30]

Running through much of this type of initiative is growing awareness of the 'supply' or 'value' chain associated with the smallholder farmer. The problem is that simply focusing on boosting subsistence farming through new technology will not adequately promote sustainable development for this class of economic activity. And while science has opened up enormous possibilities for farming yields, these are often simply not being exploited. The key issue is how to integrate the smallholder farmer downstream into better market access and upstream into better input provision. And this is an institutional issue since the types of thinking, infrastructure and organisational conditions at present obtaining often do not allow this easily to take place. There is, we believe, no silver bullet here that provides a ready-made solution. Rather the search for solutions is an experimental one. But it is one that must engage both upstream and downstream across the total supply chain in an integrated way.[31] Agricultural scientific research is an essential part of this but it too must engage appropriately.

Summary and conclusions

It is within this broad context that DFID established the RIU initiative. It is doubtful if the Collier (2008) view has much traction since wholesale reliance on large scale agriculture is impractical as a continental solution given the wider implications for urban poverty and unemployment.[32] On the other hand agricultural science can no longer be left on its own to meet the new demands of the twenty-first century using old institutional methodologies. There is still a need for science of course, but it is becoming clear that it should be informed by and channelled to the needs of client sectors to a much greater degree than has been the case in the past. In turn this means new types of relationships with other stakeholders and new types of capacity on the part of scientific institutions and organisations. This does not mean any reduction in the quality of the science. Rather the reverse in fact, as May (2004) has pointed out. It implies that scientists and the organisations in which they work, need to improve their capacities to undertake quality science. But to do this they must also become more aware of the socio-economic context of their research and how this can inform the nature and purpose of what they are trying to do.

This chapter has been written to provide a broad contextual background to the RIU programme. Its main thrust has been to show how traditional approaches to the production and use of scientific knowledge in agricultural development on very poor countries have been based very much on conditions obtaining some 50 or so years ago, those enshrined in the Green Revolution. However, times have changed. Development agriculture is much more complex in many respects and the persistence and worsening of agricultural poverty in many parts of the developing world suggests the need for new thinking. The RIU programme has been an attempt at just this. The Best Bets have been technology development projects funded by technical assistance aid (DFID) and operated by consortia which have aimed to mobilise private sector inputs and stimulate small-scale entrepreneurship.

The monograph is divided into seven chapters. Chapter 2 provides an analysis of cognate science policy that stresses issues of resource allocation to research, the knowledge creation/validation process, its impact on innovation and its particular problems in the context of low income agriculture. Chapter 3 outlines the RIU itself, showing how the programme has been an institutional experiment in which progress has followed an evolutionary path subject to periodic review and reformulation. This chapter ends with a description of the Best Bet sub-programme, its rationale, planning, funding and inception. The following three chapters cover all nine of the Best Bets that were eventually supported. They summarise the general problem areas, the research projects that had provided the relevant knowledge, practical aims and objectives, and progress to date.

The cut-off date for monograph purposes had been set up as December 2011 though some (but not all) projects have continued to receive support beyond this date. Detailed sources have been sourced through programme and M&E documentation, supplemented by interview and internal management materials.[33] Chapter 7 concludes the monograph. It analyses what the experience of the Best Bet initiative of the RIU can tell us about putting research into use and what policy implications this has for future technology assistance to low income countries (LICs). It focuses on patterns of technology use, mobilisation of relevant private sector activity, capacity building (particularly for local tertiary education and research bodies), innovation systems and the role of national governments. It finally provides examples of new private sector financial interest in some of the projects.

Notes

1 There was one South Asian project but this has been omitted.
2 Though as Juma (2011: 8) and Conway and Waage (2010) point out, the impact in South Asia has been better.
3 Although it does suggest by implication that their sheer scale and complexity do not on the whole provide 'value for money'.
4 The Comprehensive Africa Agriculture Development Programme (CAADP) was established under the NEPAD (New Partnership for African Development)

in 2003. CAADP's Pillar 4 deals with Agricultural Research, Technology Dissemination and Adoption. See www.nepad.org/system/files/caadp.pdf, p 1.

5 The literature here is enormous. In addition to other texts mentioned, reference may also be had to Conway (2012), early chapters. See also the World Bank *Development Report* (2008) and related documents.

6 See for example Nicholas (1971), Chapter 1 for a discussion of this point.

7 Clark (1985) discusses this in some detail. See Chapters 7 and 8.

8 See for example Solow (1957).

9 See Smith (2009) for a detailed discussion of this point. See also Marglin (1996); Esteva (1996); Perkins (1997); Cullather (2004); all cited in Smith (2009).

10 Perkins (1997) refers to the resulting narrative as the Population-National Security Theory (PNST), whereby overpopulation was argued to lead to the depletion of natural resources, which would consequently give rise to severe food shortages or even famines. These, in turn, would instigate political instability thereby paving the way for communist insurrection, which would eventually jeopardise American interests abroad. See Smith (2009: 46).

11 See Conway and Waage (2010) Chapter 5, p. 119.

12 See CGIAR (1992), p. 4.

13 Again, there is an enormous literature on these aspects. See for example Chambers & Ghildyal (1985), pp. 1–30 for an early critique and more recently Hall (2009).

14 Reference in particular should be made to Task Force 10 on Science, Technology and Innovation; See Juma and Yee-Cheong (2005).

15 See May (2004). Similar sentiments are expressed in a contemporary DFID Select Committee Report. See DFID (2004).

16 There is inevitably some overlap among these points.

17 However, it is still the case that many donors are only slowly getting involved with science policy issues as such. Many still focus on conventional funding of science projects. For a discussion of the UK position on this see the Report of the House of Commons Science and Technology Committee on the Use of Science in UK International Development Policy. In response DFID appointed a Chief Scientist to help fulfil a more targeted developmental function. See DFID (2004).

18 See Conway and Waage (2010, chapter 8) and Conway (2012, chapter 15) for a summary of climate change issues and possible responses.

19 See Mugagwa, Wamae & Outram (2010) for a comprehensive discussion of this and related issues.

20 See Juma and Serageldin (2007), Clark *et al.* (2002), Clark *et al.* (2007). An up-to-date summary of biotechnology and its potential role for small scale agriculture is provided in Conway (2012), chapter 9.

21 This point will be developed later in the text but see Anderson (2008).

22 See Wiskerke and van der Ploeg (2004).

23 Abundant examples of this methods-driven debate can be found in the general literature. For critics see Tripp (1989); Biggs (1990); Biggs and Smith (1998); and Hall and Nahdy (1999).

24 See also Hornberger *et al.* (2007).

25 FAO (2011), p. iv.

26 All emphasise the need to better integrate smallholder agriculture with markets. An essential component in this view is the need to aggregate effective market demand through schemes such as warehouse receipt storage systems and contract farming schemes.

27 See SFSA (2011). We are grateful to George Osure for providing information on these activities.

28 See SFSA (2011) p. 10 onwards for further details on the OAF.
29 Interview data.
30 Another interesting case is that of Genesis Farms in northern Sierra Leone. Its aim is to introduce cash crops for Sierra Leone, which can be produced on a commercial basis, with production levels beyond the domestic needs of rural families. Excess production will provide income to rural families and substitute imports of staple foods. See http://teruginmakomp.blogspot.co.uk/p/genesis-farms-sierra-leone-ltd.html. We are grateful to Frances Kimmins (Agriculture and Rural Livelihoods) for drawing this example to our attention.
31 See, for example, Horton *et al.* (2010) for a review of an approach to this issue. Interestingly a range of new NGOs are appearing that promote integration of smallholder agriculture across complete value chains. A good example is CABE, whose mission is to build linkages and entrepreneurial capacity of smallholder farmers and youth by addressing the disconnect between local practitioners and policy makers. We are grateful to Hannington Odame for drawing this example to our attention.
32 The Collier position really avoids the issue of how this is to be achieved within a reasonable time frame. It took Europe through enclosures and early industrialisation some 200 years. Stalin managed it in a decade but with appalling social costs in a highly centralised economic system. In Africa, however, while there is of course some large scale farming this impinges on a small minority of the population. The only way forward is to work within existing economic structures and look for appropriate institutional change.
33 And for simplicity and ease of exposition, references to specific points are not generally provided in the text. Further details can be found in the supporting RIU discussion papers cited in Appendix 1.

2 Science policy and the knowledge market

Introduction

In the last chapter we showed how practical approaches to technology development in poor country agriculture are still dependent on the Green Revolution model. And in the context of extensive economic changes over the past half century there is a *prima facie* case to be made for changes in relevant science policy. In this chapter we shall expand on that argument through direct engagement with science policy analysis. Central to the argument will be the proposition that the 'market' for knowledge is asymmetric. While the production and validation of knowledge has a determinate 'supply price' (based largely on its cost of production) there is no equivalent unique 'demand price' since the value of any piece of information is entirely dependent on its context of application. This means that any knowledge market (unlike a product market) can never clear and can therefore never allocate resources optimally except in the special case where proprietary rights get established (e.g. through the patent system). In turn this means that it is entirely possible, indeed normal, for considerable resources to be spent inefficiently.

This is not an original proposition. It returns to a debate within the UK that has been argued for many years going back to the writings of eminent scientists (such as Haldane and Blackett) in the pre-war period. Important examples of this debate can be seen in C P Snow's Reith lectures in the late 1950s,[1] the publication of the UK Rothschild Commission report in the early 1970s and the establishment of a series of UK government initiatives in more recent years.[2] In every case, the attempt may be seen as one of improving the operation of the knowledge market and to economise on scarce public resources. This has never been such a problem in the private sector since firms are better able to restrict access to proprietary information, as they must do if they are to stay in business. They are also able to integrate knowledge across the whole value chain as Bell (2007) has pointed out. Arguably the issue has taken on more importance in recent years as knowledge has become more globalised and economic production has become more dependent on technological change. There are still many unresolved policy issues.

The first section defines science policy analysis showing that it is basically a branch of applied economics, albeit one fraught with unusual complexity. The next section explores briefly attempts to broaden analytical content by focusing more on innovation and on the systemic context within which innovation may be better promoted. This so-called *Mode 2* position has achieved some importance in the academic literature[3] but has not so far affected the practicalities of science policy in LIC agriculture. One suggested reason for this may be the complexities involved in linking research funding to an evolving and uncertain institutional context. The following section suggests an alternative approach that treats knowledge as a commodity, albeit one with special features. The analysis goes into some detail on the 'knowledge market', explaining why scientific knowledge is essentially a form of public good, but one that is inherently difficult to apply efficiently to economic production. It also argues that its successful operation is closely linked to the systemic context in which it is embedded, a position quite close to that of *Mode 2*. The final section summarises the generic argument and shows how it relates to the RIU programme.

Science policy

It is commonplace nowadays to argue that in some general sense technological changes have had a fundamental impact upon economic growth since the industrial revolution. However, the form this impact has taken and the social relationships involved are even now understood only very imperfectly. In particular, the influence of social expenditures upon science (through R & D, scientific institutions etc.) is problematic. There is no question that there have been significant economic breakthroughs involving systematic scientific research, but technology development and productivity advance usually also occur as a result of organisational changes, the migration of skilled manpower, induced changes, learning, luck and a variety of other social mechanisms and combinations of thereof.

Nevertheless, expenditures upon organised science have increased dramatically over the past 50 years or so and much of this has been funded by governments to be spent through their own agencies/ministries etc. or externally through private bodies, mainly industrial firms. Clearly, such high and growing levels of expenditures have involved governments in a resource allocation problem. If massive sums of public money are to be spent in harnessing science and technology to the national interest then criteria are required which will permit the disposition of such sums on a rational, and it is to be hoped, socially optimal, basis. Typical policy questions are:

- How far should governments fund private sector research?
- What mechanisms should they adopt to this end (e.g. direct subsidy, tax relief, provision of special facilities, public purchase)?

- What sectors should be given priority?
- What areas of basic science research should be funded by governments, in what proportions and through which institutions?
- How should higher education and science policy be co-ordinated?
- How should spending ministries fund their own research?

This broad policy issue has generated an on-going debate for some considerable time in the UK, one side of which has been consistently promoted by the scientific community; that is the position of not interfering with scientific autonomy. The argument is standard. Attempts to constrain science by relating funding to social welfare considerations will endanger the 'seed corn' of future breakthroughs. Scientific research can only be assessed by the scientists because they and they alone, fully understand the significance of what they do. Funding decisions should therefore be made on the basis of peer review organised under the aegis of academic disciplines. There may of course be general resource constraints at a macro-economic level but once governments decide on overall funding levels for science, the scientific community should be left to run things itself. Otherwise, it is argued, the quality of science will be compromised.

In support of this position, examples are given of interference that has gone wrong. One famous example is the alienation of Medical Research Council (MRC) funds to the UK Department of Health and Social Security (DHSS) in the 1970s, a decision prompted by the Rothschild Report (Rothschild, 1971) but quickly running out of steam as the DHSS had problems acting as a 'customer' to the MRC's new 'contractor' role.[4] The Rothschild Report was essentially an attempt to rationalise science expenditures that were threatening to run out of control and was intended as a means of establishing a kind of market for science and technology expenditures through which the customer, a government spending department, would identify research needs arising, or likely to arise, from its normal activities and request the contractor (which would often be one of its own laboratories, but not necessarily so) to carry out the relevant research. The hope was that by establishing a form of market mechanism, government laboratories would be disciplined into social relevance; or at least things would be moved in that direction and resources conserved.

Quite clearly the Rothschild reforms, and the type of thinking that underlay them, raised questions about the economic content of science policy decisions. In particular, they show how such decisions are economic in the sense that they are usually concerned with questions of resource allocation under conditions of scarcity. These are important questions since the decision to allocate £X to function A_1 means that this £X will no longer be available to support functions $A_2 \dots A_n$; i.e. function A_1 has an opportunity cost, namely that of *not* supporting other possible functions or projects. It follows that the committal of resources to science and technology projects represents a form of social investment in the sense of adding to the available stock of capital and as such, at least in principle, it can be handled analytically in much the same way as any other form of investment appraisal, i.e. as a problem in constrained optimisation.[5]

Unhappily there are many factors that make this a difficult task. Future costs and benefits are hard to compute with any certainty, especially when they involve placing a valuation on research outcomes which have yet to be realised in commercial terms. For these and other reasons the application of a simple customer/contractor mechanism is a rather blunt policy instrument. It is factors such as this that have made the 'seed corn' argument a hard one to argue against in practice. However, in spite of this, funding bodies have continued to introduce institutional innovations designed to constrain science expenditures. An early example of this was the establishment of the UK Biotechnology Directorate, a body whose remit was to ensure that the UK could better exploit relevant research in the biosciences, in particular by linking research council funding more directly to industrial practice. But unlike the application of a *simpliste* customer/contractor principle, the Directorate concentrated more on organisational changes that would encourage greater dialogue between science and industry.

It is useful to expand a little on the Directorate case since it is an early example of putting research into use, or rather to deal with the relative failure of UK institutions to do just that. Throughout the 1970s it had become increasingly obvious that, no matter how successful British laboratories had been in purely scientific terms, the fruits of this success were enjoyed largely by foreign-owned firms. Sharp (1989, 1996) for example, shows how the discoveries associated with monoclonal antibodies (in MRC laboratories) were patented and exploited largely by US-based companies.[6] And it was issues of this type that led to the setting up of a commission to determine what might be done in the case of the biological sciences. The Spinks Report was published in 1980. It made a series of recommendations that included the establishment of an industrial enterprise (*Celltech)* and an innovative technology directorate designed to promote pre-competitive research in the development of therapeutic proteins (the *Protein Engineering Club-PEC*).

Essentially, the PEC was a device designed to target research council support with complementary financial input from private companies. Figure 2.1 illustrates the structure. The Biological Sciences Committee of the Science and Engineering Research Council (SERC)[7] funded a four-year research programme to the tune of some $5 million; 15 per cent of this came from four companies, the remainder from the SERC. The primary objective was one of bringing together interdisciplinary research groups to work on the building of therapeutic proteins, to design new technologies and instrumentation, and to establish a cognate database. The programme funded some 23 projects at 11 universities and provided 36 post-doctoral fellowships. The active involvement of an economic interest was enhanced also by having the PEC Steering Committee chaired by an industrialist. Sharp goes on to show that the venture not only merged scientific and economic stakeholders but in addition provided a mechanism to integrate disciplines across science and engineering in a way seldom previously achieved. Indeed such was the success of the PEC that at the end of the programme period all four companies indicated that they were now convinced of the technology's viability and would therefore no longer wish to contribute to a collective venture. The PEC was then discontinued.

From an RIU standpoint what is significant here is the use of public resources to fund the riskier aspects of technology development. Protein engineering required the combined inputs of X-ray crystallography with molecular biology and computer science, disciplines that would not normally interact. No entrepreneurs were willing to fund such a non-proven activity even though there had been some indication of success in Japan.[8] Instead public money filled the gap. Nor was there any issue raised about 'seed corn' being harmed by commercially driven science policy. We shall see later on that the RIU Best Bet programme has performed a remarkably similar function, not in this case, involving cutting edge science, but rather by covering the tricky problems of promoting technology development and entrepreneurship in a LIC setting.

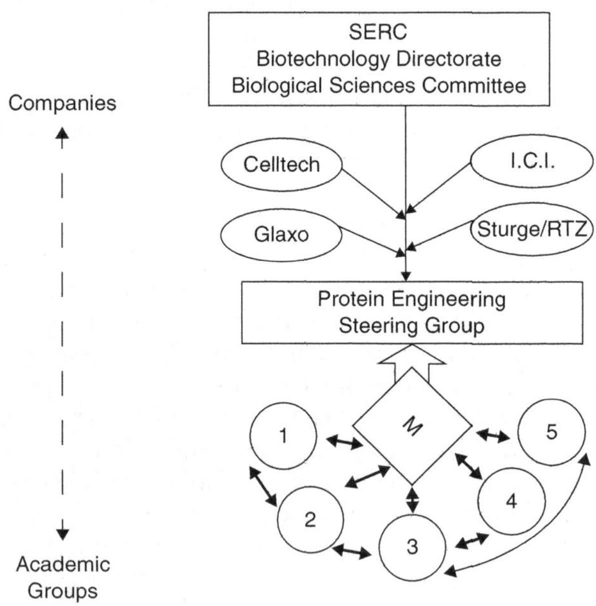

Source: Sharp (1996), p225

Figure 2.1 Protein Engineering Club

Innovation

The Protein Engineering Club is only an example of course. It has since been followed up by other similar measures of public support like the DTI Link Programme, and the OST Technology Foresight Programme. But it is examples like this that have begun to open up new perspectives on science policy discussion. Still one of the most persuasive treatments of this was *The New Production of Knowledge,* the book published by Michael Gibbons and colleagues in 1994. The essence of the Gibbons position is that policy focus should shift from 'science' to 'innovation' since it is innovation that underlies welfare improvements. Investment in science is still important of course, but experimental knowledge is only one form of knowledge. It needs to be complemented by many other forms necessary for economic production to take place and welfare enhanced. And such ancillary knowledge is not normally held in the laboratory but is rather more tacitly found in the experience and capacities of many different types of people and organisations. Moreover the totality of this knowledge is systemic in that it is normally a combination that is needed for innovation.

According to Gibbons therefore there is a strong case to be made for organisational changes that bridge the conventional divide between knowledge search and knowledge use. What he labelled the *Mode 1* position does not do this since this was based on a strict separation between the two activities. Rather it promotes a division of labour that confines knowledge search (and validation) to specialised laboratories and discourages so-called 'downstream' linkages. For Gibbons this approach to science policy inhibits innovation since it breaks up the necessary organic connectedness that allows innovation to succeed.[9] And since much modern policy is designed to promote such systemic connectedness what is needed is a policy position (*Mode 2*) that legitimates a more connected role for scientific bodies. Henceforth, he and his colleagues argue that science funding should reward proposals that encourage such behaviour.

Support for this position is growing. For example, a recent publication (Wooding *et al.*, 2011) on cardiovascular research has concluded that on the basis of 29 grants studied, only a small proportion of these had wider impacts although all fulfilled traditional output criteria such as publication and research capacity building. The authors argue that more attention should be paid to wider societal benefits in the funding of medical research and (by implication) that there is no necessary conflict here with good scientific practice. Nevertheless, there is still reluctance on the part of scientific communities to engage with the wider context of their experimental work.

Arguably one reason here is that scientific communities have difficulty engaging with policy issues simply because they have little background in social science. The trained social scientist is someone who has built up a certain understanding of a given part of the social universe, who identifies problems which he/she believes to be important (or who is commissioned to analyse problems that the sponsor regard to be important), who then tries to analyse the various social forces which have interacted to bring about these problems and to identify the major causal factors. He/she will then use the analysis to prescribe possible remedial policies and to suggest what the implications of these might be. However, since the world is a very complex place and since he/she will not in any case be responsible for policy execution, the social scientist ought not in general to be too categorical about the advice given and should if possible anticipate the consequences of certain courses of policy action.

What then does this sort of activity imply? Clearly, an important implication is that there is no direct recourse to any unified theory which will instruct the analyst as to how to proceed. This is partly because of the inherent nature of policy analysis as described above, but it is also due to the fact that most social problems are by definition interdisciplinary, and those that concern science and technology policy are very definitely so. Thus their analysis requires some elements of sociological understanding because, for example, it is necessary to understand certain features of the behaviour of scientific and industrial communities. They require similar expertise in political science because there are important power/institutional dimensions to science/technology policy decisions; similarly, as we have seen, with economics. Usually also some detailed natural science or engineering knowledge is required because there is a need to know about the scientific activity in question or the nature of the productive activity under investigation. There is little point, for example, in studying the impact of monoclonal antibodies if one is not willing to learn about what forms they take and how they are developed and used in more complex health systems.

The analyst is forced into drawing from a wide range of concepts and postulated relationships and herein, of course, lies a major difficulty since academic work is traditionally structured along disciplinary lines. Things have developed this way because it has proved the most efficient (in an academic sense) method of extending the boundaries of knowledge (the research function) and to communicate a coherent body of ideas to students (the teaching function). The discipline then determines the way that scholars think, the problems they regard as important, the language with which they communicate, the research techniques they adopt and, ultimately, the criteria they use to judge the performance of their peers and to determine the *ex ante* acceptability of research projects.

These features then become enshrined in more general bureaucratic forms including budgetary procedures, levels of decision-making within academic institutions and grant-awarding bodies, and career structures for staff. Hence knowledge about the world is pursued and diffused in ways which are not in general consonant with practical problems faced by people and institutions who have to make decisions on matters of public policy or, indeed, people who are concerned with economic production. In a very general sense, therefore, there is a problem of academic alienation and the science policy analyst is often faced with the difficulty of having to know enough about the different disciplines so as to be able to use them in the evaluation of a given issue.

Knowledge market

An alternative approach to the *Mode 2* argument (and one that might better focus the attention of decision makers) is to focus directly on 'knowledge' as a commodity bought and sold in a market. The idea of portraying knowledge as an economic commodity is perhaps counter-intuitive but since the production of knowledge is an activity that requires scarce resources and since these are often funded by the tax payer (they are a form of public good) there is we believe a *prima facie* case for doing so. In order to give better analytic content to the idea of a knowledge market it is useful to make a distinction between 'information' and 'knowledge'. In fact conventional economic analysis does not do this but on the whole uses both concepts interchangeably to describe a category of boundary conditions that prevent economic actors from behaving as they ought to (to achieve some optimum in terms of the allocation of resources).[10] What the formal theory of information does is to define the concept in a way that is logically independent of 'meaning' and in so doing provides the basis for a more general theory of systemic communication.

This point has been explored in some detail in Clark and Juma (2013: chapter 5) and is worth exploring briefly as it allows us to develop a model of technology development that has the property of systemic generality. Classical information theory conceptualises 'information' in terms of a flow of 'messages' which have 'news value'; that is, they cause surprise to recipients. In the words of Singh (1966: 9) for the 'communications engineer's purification of the term, the stress is on the quantitative aspects of the *flow* in a *network* of an *intangible* attribute called *information*'. Such a network consists of three main parts:

- a transmitter
- a receiver
- a communications channel.

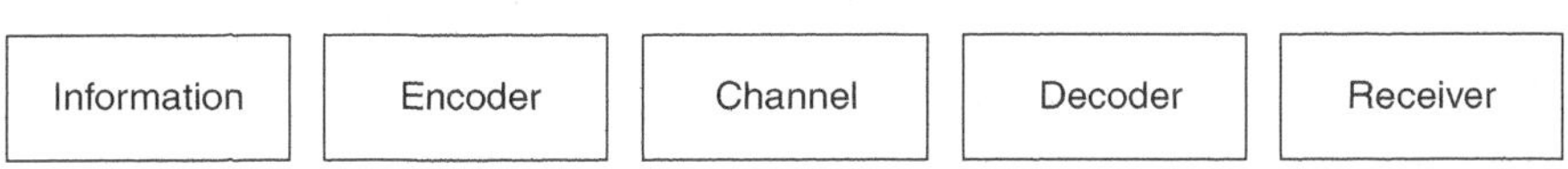

Source: Author's diagram

Figure 2.2 A simple communications system

Since the medium of communication is not in general similar to that of the transmitter or receiver, there needs to be a means of translating signals at the beginning and end of the process, so that any typical communication system may be represented as shown in Figure 2.2. For example, a traditional radio system operates by converting sequences of voice production into electromagnetic waves that are beamed through the atmosphere to be received and understood by listeners. Thus 'information' turns out to be closely related to energy. Its flows are conventionally quantified in statistical terms, that is by means of a multiplicand of probabilities attached to 'bits' of information where the total information content of any message, in a given 'ensemble' of possible messages, may be measured by the logarithm of the probability of its occurrence. Logarithms are used to convert combinations of discrete probability measures to a summation. Hence the information content in any complex of messages is then simply the sum of its individual components. Singh (1966) shows that no information system has this freedom of information content because of practical limitations. The English language, for example, is such that not all of its letters are equally likely to be chosen. It has the property of redundancy and it is this redundancy that gives it intelligibility.

It may easily be seen also that redundancy is closely related to organisation. This is because as soon as constraints are placed on a system (in this example, to give language intelligibility), to that extent the system exhibits a degree of organisation that it would not have had in the absence of such constraints. Notice, finally however, that the above definition is completely independent of semantic meaning. It is merely 'a measure of one's freedom of choice when one selects a message from the available set, many of which may well be devoid of meaning'. It is thus used logically in computer science. The greater the amount of raw information in any given system the more chaotic that system will become and the greater the need for internal organisation to give its component parts meaning. It is in this very fundamental sense that organisation is so central to the idea of knowledge use. Unless the institutional context is such that information flows have meaning it will not function as a dynamical system and will not in consequence contribute significantly to economic growth and development.

Two properties immediately follow from this discussion. The *first* is that information only becomes useful knowledge if the receiver perceives it to be so. In a sense, the receiver, whether an individual or an organisation, needs to understand information as economically productive. To the extent that understanding is not there, then the information loses value. From an organisational standpoint the capacity to convert information into useful knowledge, and thence economic output, is therefore contingent on institutional conditions. Put differently, unlike economic commodity flows, information flows only take on value where there is a deep and shared understanding of what that information means on the part of both the sender and the recipient.[11] And it is this peculiar property that renders the market for knowledge so hard to function efficiently. For example, it is one important reason why centralised public sector laboratories often find it hard to interest the private sector in their research findings. The *second* property is that productive information flows are also therefore a function of related actions on the part of recipients. This is because the capacity to understand the full potential of externally acquired knowledge is governed to a large degree by recipient knowledge investment. Paradoxically technology transfer in general only works well where the recipient carries out its own related research programme.

These relationships are not only technical but are also ideological since in a fundamental sense, organisations only hear what they want to hear. For example, contemporary philosophy of science now accepts that all knowledge is contingent knowledge in the sense that its results are theory dependent. The idea of the 'paradigm' popularised by Toulmin, Kuhn, Lakatos and others in the 1960s[12] was originally developed to show how scientific progress is strongly influenced by institutional views about the true underlying nature of reality. And it was not long before the idea of the 'technological paradigm' was used to define an established technology that guides production. In effect it supplies the *informational* coherence that provides for the knowledge market what the product itself provides for the product market. Notice also that the paradigm crosses established institutional boundaries.[13]

There is one additional point that should be stressed about this analysis of information theory. That is, it illustrates by implication the need for well-organised and co-operative knowledge markets. Nowadays (and partly also because of technological changes in information handling) there is so much information that could be relevant to economic agents that new ways have to be found of minimising the costs of information search and maximising the 'receptivity' of those who are able to use the information. It is this essentially interactive feature that ultimately lies behind effective science policy. A context which permits information interchange among otherwise independent organisations, where management hierarchies do not intrude excessively and where encouragement is provided to individual agents to try out new and possibly risky approaches (and even to make mistakes), stands every chance of contributing greatly to innovation and change.

Conversely, a system that is formal and rule-bound, where individual agents are constrained by their own hierarchies and where co-operation between organisations is viewed with resentment and suspicion, is unlikely to make that sort of positive socio-economic impact. It is with this disjunction in mind that many industrialised countries, driven to some extent also by globalised competition, have begun to experiment with new and innovative institutional forms of the emergent knowledge market. Notice finally that such knowledge channels are not simply those associated directly with formal scientific output. What is equally needed are channels of communication (or networks) that build awareness of alternative technology sources, public policy regimes that can help or hinder transfer processes and incentive structures that permit co-operative arrangements among organisations that currently treat each other as rivals rather than as potential sources of new knowledge. Arguably it is this essentially interactive property that underlies much of contemporary innovation systems analysis.

Summary and conclusions

Let us summarise the chapter's argument as follows:

1. Science and technology policy analysis is still a relatively unexplored area of public policy discourse. It may be defined very broadly in terms of how and why social units commit resources to science and technology, what sorts of problems arise in so doing and what sorts of improvements might be made. It has become increasingly important because of its resource implications in modern knowledge-based economic systems.
2. Much of the reason for its development depends upon demands on the part of governments for expert assistance in the making and monitoring of policy, including how to fund scientific organisations. Science policy should therefore be regarded as a form of social policy and its analysis akin to that of the social sciences rather than to that of the natural sciences. It also means that analysis is problem-focused and interdisciplinary in the sense that no one academic discipline by itself can provide the necessary conceptual background, although normally the examination of any issue will require complementary input from the natural sciences.[14]
3. The role of economic analysis is central to science policy analysis partly because at bottom, the issues involve the allocation of scarce resources, partly because much relevant theoretical discussion uses economic conceptualisation and partly because the analytic technique, if correctly employed, provides a useful method of sorting out the complexities of any given set of issues.
4. However, what renders economic analysis especially difficult is the asymmetric nature of the knowledge market. Because demand price is indeterminate the value of any piece of information is entirely dependent on the context of use. For this reason the knowledge market cannot easily clear unless the knowledge involved can be privatised, as with the award of a patent for example. And it is therefore entirely possible for considerable sums of public money to be wasted.
5. Various attempts have recently been made to rationalise the analysis of the knowledge market through focusing attention on innovation as a

systemic phenomenon. Here the literature has begun to focus on scientific expenditures as but one (albeit important) part of innovation while understanding the wider context of application is equally important.

What might all this mean for technology development aid in LICs? In our view it suggests an agenda in which policy analysts will become much more aware of scarce resources and wasteful expenditures. In such a context, scientists should expect to pay much greater attention to the wider context of their work than has traditionally been expected. The biosciences, for example, tend to have the health and agricultural sectors as their most immediate clients. And impacts such as health improvements and food security tend to be high on the agenda. Accordingly, we might expect bodies funding the biosciences to put pressure on grantees to show wider welfare results at minimum cost and to demand evidence that resources are not used wastefully (for example, rediscovering what is already known).

In turn this will require scientists and their institutions to link more closely to their downstream systemic partners and to be able to better integrate knowledge across organisational boundaries. Such integration between 'knowledge search and validation' and 'knowledge use' is common in much of industry as Bell (2006; 2007) explores in some detail. Bell's argument is reached on the basis of detailed empirical research over the period 1950–2000. It shows that successful technology development is largely enterprise-based and relies on 'public sector science' only to a limited extent. R&D is important of course, but it is not where innovation mainly takes place and it is innovation, not R&D, that really drives possibilities for poverty reduction. The mistake was made in the early post-war years at a time when development was seen as the responsibility of the public sector and technology was assumed to be entirely science-led. It seemed relatively straightforward for bodies like UNESCO to advocate the creation in poor countries of organisations (like universities for example) that were in effect 'cardboard cut-outs' of their rich country parents. Relevant technologies would thereby 'take root and flourish' and poverty would inevitably disappear. Now we are not so sure.

But Bell's position is clear. It really does not matter whether you are considering a cassava processing plant in Ghana or a deep water petroleum facility off the coast of Angola. In either case the investment activities associated with any new venture will follow roughly the same rules. The firm will determine the macroeconomic and government regulatory context; it will specify the process and product design, the ancillary facilities such as power and water supply, the necessary financial and due diligence components, and associated contracting and sub-contracting arrangements for its engineering. Management of the package is a highly skill-intensive process and one that takes time. It will of course hopefully embed the latest knowledge as a necessary condition but in practice every project is a new project and it is in this process of 'getting it right' that much of the necessary learning and innovation takes place.

Bell also shows that in general the resources needed here are many multiples of basic and applied 'research rich countries costs'. It is in the *doing of it* that knowledge is expanded. And this is where the private sector is so successful. No enterprise would tolerate the levels of economic inefficiency routinely exhibited by public sector science. It could not afford to. Though not explicit in its early inception this argument was certainly a factor in the thinking that led DFID to establish the RIU programme. While recognising that RNRRS research projects supported over the 1995–2005 period were geared towards specific technology development problems, it had by then become clear that a fresh approach was needed to ensure continuous interaction between knowledge search and its application. The RIU therefore decided on a twin-track policy of exploring how effectively the knowledge produced in the RNRRS period had been used and where there were problems, how these might be mitigated in future periods. The next chapter summarises how this was done.

Notes

1 See Snow (1963).
2 For example Technology Foresight, the DTI Link Scheme, the Biotechnology Directorate, the Teaching Company Scheme (as of 2003 re-named as Knowledge Transfer Partnerships).
3 The best original source for this is still Gibbons *et al.* (1994). See also Nowotny *et al.* (2003).
4 See Gummett (1980) for a detailed case study.
5 Sometimes this is couched in terms of a 'rate of return' to investment, a concept that is properly used in a micro economic sense holding other inputs constant. Unfortunately there is frequent usage of this idea in a much broader way to estimate rates of return in a macro economic sense over long time periods where other inputs cannot be held constant.
6 See also Bud (2010).
7 In 1994 the SERC was split into the Particle Physics and Astronomy Research Council (PPARC), the Engineering and Physical Sciences Research Council (EPSRC) and the Biotechnology and Biological Sciences Research Council (BBSRC).
8 See Fransman (1995) who describes the use of MITI to develop the Protein Engineering Research Institute at Osaka University. See also www.protein.osaka-u.ac.jp.
9 And indeed there is the related argument suggesting that better engagement with wider contexts will improve the quality of the science itself.
10 There is a long history of the conflation of information with knowledge going back to Machlup in the 1950s at least. See Machlup (1962). This discussion has been developed in more detail in Clark (2002) and Clark and Juma (2013).
11 There is of course an informational content in any normal economic transaction but outside the case of very complex goods we can probably treat this as fairly trivial. For those who cannot accept such treatment, a more general position could be to define pure 'economic' transactions as having no

informational uncertainty, but where they do, to stress the contingent nature of such transactions in a purely informational sense.

12 See, for example, Lakatos and Musgrave (1972).

13 The notion of the scientific paradigm stemmed originally from Kuhn (1970). For an account of its later use in technology policy see Clark (1987).

14 For a detailed analysis of interdisciplinary approaches to science and policy see Lyall *et al.* (2011).

3 Research into use programme

Background

While the first two chapters have provided a generic background to the RIU, this chapter will summarise the factors that gave rise to its inception as a direct outcome of the DFID RNRRS which ended in 2005. It will summarise how and why the programme radically shifted in emphasis halfway through its operations leading to a new business plan, management structure and log frame. It will show how the Best Bets programme was intended to fit into the wider set of RIU activities and summarise the procedures through which the Best Bet projects were selected through a 'dragons den' in November 2009.

In 2006, DFID published its third white paper, setting out the UK government's policies for eliminating poverty worldwide. The white paper was preceded in 2005 by DFID's Agriculture Policy Paper on Growth and Poverty Reduction which focused on promoting growth in this all-important sector through the spread of new technologies. This paper also outlined DFID's commitment to enhancing the resilience of farming households to external shocks such as drought or disease, which can plunge already vulnerable households into deeper poverty. It emphasised the need to improve access of poor people to knowledge and technology, through both public- and private-sector institutions. In the same year DFID's Central Research Department (CRD) published its Research Funding Framework (2005–2007) which identified agriculture as one of its priorities. Subsequently, a Strategy for Research on Sustainable Agriculture (SRSA) was prepared. The Research into Use (RIU) programme became the first to be developed under this new strategy. It proposed to contribute to this strategy by adopting a pro-poor innovation systems approach to getting research into use and to increase the understanding of how this is done.

The research that would be 'put into use' derived from projects funded under DFID's Natural Resources Research Strategy (RNRRS), which consisted of some 1,600 projects running from 1995 to 2005. The focus of the RNRRS had been to improve the livelihoods of the poor through better management of natural resources. The ten research programmes launched under the strategy were designed to generate new knowledge and promote its uptake and application. They addressed the needs of people living in a range of agro-ecologies including semi-arid areas, high-potential areas, highlands and tropical moist forests; and those at the forest/farm, land/water and rural/urban interfaces. The breadth of projects reflected the multiple routes by which research can have an impact on poverty.

The RNRRS had seen significant evolution over its life (DFID, 2005). This included a shift in focus from generating research and producing scientific publications to emphasising the impact of research on poverty. The focus also moved from outputs to outcomes and long-term impacts. At the same time, interdisciplinary research, the policy environment and the livelihoods of the poor began to receive greater attention. One of the most influential legacies of the RNRRS was the use of innovation system principles in the development of new partnerships, products, processes, markets, institutions and organisations that are better equipped to put research into use.

In the context of RIU, innovation meant the use of new ideas, new technologies or new ways (processes) of doing things in a place or by people where they have not been used before. Of course, innovations in this sense have been constantly taking place through the actions of rural communities themselves, but the intense pace of global change and the threats from climate change and environmental degradation means the poor in developing countries must rapidly adapt just to cope. Innovation, meaning the use of new knowledge in a given context, often involves working with and re-working the existing stock of knowledge (research into use). It is often the key to building better and more sustainable livelihoods, because new knowledge is required to deal with the rapidly changing environments that face farmers and other rural people. It often involves local creative imitation and adaptation, rather than the development of something radically new. It is usually achieved through many small improvements (e.g. in production technologies, processing and institutions) rather than through a few big sweeping changes. And it involves greater ownership of the process by poor people themselves.

The RIU approach was therefore to shift the focus of attention away from the important tasks involved in the generation of new knowledge to the ways in which that knowledge can be put to productive use. An innovation system is usually seen as a network of organisations and individuals involved in generating, modifying and using new knowledge. The networks might be national, sub-national, regional or international. They comprise not only the users of the knowledge (farmers, consumers, artisans, labourers and traders) and the producers of new knowledge (researchers) but a host of intermediary organisations including extension workers, NGOs, enterprises in the supply chain, credit agencies and government. This systems approach considers not only the totality of the entire research, development and extension spectrum, but also the institutions, systems of production and social relations in which these activities take place.[1]

In practice, the RIU was focused on South Asia and Sub-Saharan Africa (SSA). After a short inception phase it started by carrying out a series of country assessments to match local demand in selected countries to the supply of RNRRS derived technologies. These assessments were conducted in selected countries across SSA and South Asia and led to the establishment of interventions designed to promote three linked objectives:

- *Output 1* (£16.5 million): putting research into use (*Components*: improving the access to RNRRS and other research outputs [Innovation Challenge Funds]; enhancing demand for RNRRS and other research outputs [country programmes]; developing enterprises using RNRRS and other research outputs).
- *Output 2* (£8.2 million): learning about getting research into use (*Components*: monitoring and evaluation support and synthesis; impact evaluation).
- *Output 3* (£4.1 million): lessons having policy impact (*Components*: influencing the agenda of national, regional and global partners; communications with the global professional community).

Output 1 was developed as (i) a Challenge Fund operating in South Asia with 13 sub-projects; (ii) country programmes established in six Sub-Saharan African countries, primarily involving establishment of national innovation coalitions and innovation platforms; and (iii) a feasibility study of enterprise development activities involving public-private partnerships. Outputs 1 and 2 involved studying up-scaling successes in RNRRS; doing comparative research on innovation and up-scaling; promoting RIU though a website, publications and attendance at conferences. Output 3 concentrated on promoting to policy makers the need for getting research into use, mainly through the links to African regional policy processes.

There then followed a phase of establishing national teams to scale out chosen technologies and procedures for enabling policy dialogue with government and other related agencies. However, two further years into the programme it had become clear that it was not going well. A Mid-Term Review [MTR] conducted in 2008 concluded that:

> RIU (had) suffered from a range of management problems during its first two years and was currently too diffuse (with its) individual programme components – working too independently of each other. Also, due to an extended inception phase and delays in commissioning both the challenge fund and country programmes, RIU was up to a year behind where it should have been at this stage.[2]

This meant there was 'less time for the programme to achieve its objectives'. The MTR concluded that though RIU should continue, it should radically revise and simplify its operations. In response DFID changed the RIU management structure, employed a new governance team and recommended a series of revisions to the programme as a whole.

Box 3.1 Summary of reviews of the RIU programme[3]

The MTR in November 2008 (revised in January 2009), and the subsequent TR in June 2009, highlighted a number of weaknesses that had prevented the programme from delivering its stated purpose.

- The MTR found that RIU lacked a unified vision, primarily as a result of different stakeholders attaching differing levels of importance to the development and research goals of the RIU programme. While it delivered analysis of RNRRS success stories and set up strong country programmes, these programmes were of variable quality and the lack of a robust framework meant overarching lessons were not learned. As a result, the MTR recommended that RIU become more focused and decentralised.
- The TR echoed many of the findings of the MTR, and also emphasised the lack of a unified vision. It concluded that as well as over-reaching (by aiming to create national innovation platforms in a short space of time), the RIU programme was not set up to learn systematically about how to get research into use. In addition to recommending a new lean, decentralised and focused organisational structure, the TR advocated that RIU unambiguously reposition itself as a research project.

Technical review period

These revisions were based upon a technical review commissioned by the new management team. The review recommendations were then incorporated into a new business plan which was finally agreed by DFID in July 2009. The broad purpose envisaged of RIU 2 was much the same as that for RIU 1. RIU 2 would aim to 'maximize the poverty reducing impact of RNRRS and other research and by doing so significantly increase understanding of how the promotion and widespread use of research can contribute to poverty reduction and economic growth' (DFID, 2009). Being a research project, its primary aim was to accumulate, evaluate and communicate evidence on how research outputs can be delivered to its users. To achieve this purpose, RIU 2 maintained the first two outputs:

1. *Promotion → Output 1:* Enhancing the demand for and putting into use the outputs of RNRRS and other research for the benefit of the poor, for example through creating agricultural enterprises; and
2. *Learning → Output 2:* Generating evidence about how research can best be put to use, and sharing those lessons.

RIU's approach with respect to Output 1 continued to be explicitly experimental, and was intended to generate a portfolio of interventions that would need to be monitored systematically for impact, so that lessons were learned in support of Output 2 (see Figure 3.1). Integral to RIU operations were effective internal and external communications and the ability to influence the agricultural policy and research agenda. RIU would work alongside rather than replace national systems, so that its interventions were sustainable in the long run. It would play a catalytic role, creating opportunities and encouragement for uptake of research outputs. These factors were incorporated in a revised log frame developed in consultation with DFID, setting out the rationale behind DFID's investment in the programme, and its expectations regarding the programme's impact.

In reaching this goal, the RIU programme aimed to generate direct impact as well as to influence relevant policy. Direct impact would be achieved by increasing the use of innovative production regimes in crops, livestock, fisheries and forestry, increasing the average income of poor farmers, and increasing the proportion of women belonging to relevant innovation groups. The programme also aimed to foster agricultural and innovation policy changes in its six focal countries in Africa.

The link between RIU purpose and intended outputs was contingent on a number of key assumptions. At a high level, the RIU programme outputs would contribute to its purpose and goal only if the institutional and policy frameworks were in place to enable poor farmers to move to more efficient

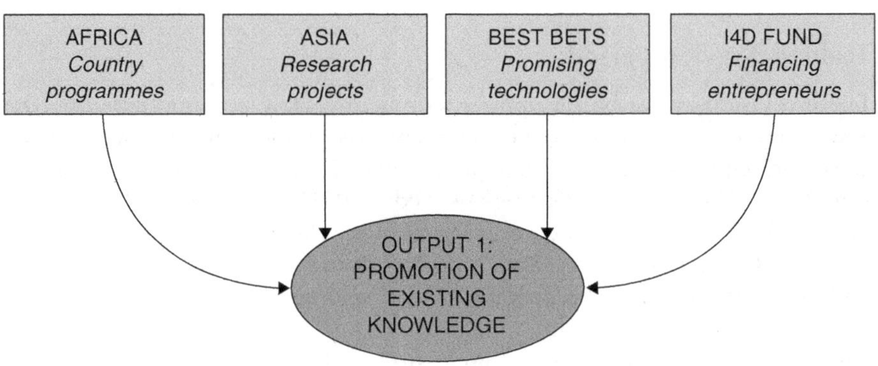

Figure 3.1 Output 1

production paths. It was recognised that in practice, making comprehensive changes to the innovation enabling environment in countries would be beyond the reach of the RIU programme alone. Rather the RIU role would be to learn about the institutional and policy arrangements that are most supportive of impact-driven innovation and learning, and to drive changes where possible. At a lower level, and within the control of RIU to a greater extent, the programme would link its outputs to its purpose. The successful promotion of existing knowledge required that RIU would be able to exercise effective leadership on a national scale, and that national policies would support innovation. Successfully learning about getting research into use thus requires an appropriate and useful framework, which RIU would develop.

Under the revised regime there were four distinct categories of interventions constituting RIU planned operations aimed at achieving Output 1:

i. six Africa country programmes focused on strengthening specific commodity value chains;
ii. a cluster of projects in Asia designed to answer specific research questions;
iii. four global best bets representing high-potential technologies to be supported directly; and
iv. Innovation for Development (I4D) fund to support SME agribusinesses arising from research outputs, through the provision of finance.

Country programmes

RIU had established programmes operating in six countries in Africa – Rwanda and Tanzania in the east, Nigeria and Sierra Leone in the west, and Malawi and Zambia in the south. Based on the recommendations of the MTR and the TR, RIU revamped its country operations to focus on those with the greatest potential impact of putting research into use to benefit the poor, as well to decentralise and empower country interventions. The two main focus countries were Nigeria and Tanzania which would receive increased funding.

Rwanda and Zambia would continue with well-performing interventions. Malawi and Sierra Leone programmes would be downsized to two activity areas closely aligned with national innovation priorities. Table 3.1 presents an overview of RIU focus initiatives in these countries, which were focused on strengthening agricultural commodity value chains.

Asia projects

In Asia, RIU would undertake a cluster of projects, drawing on its established portfolio of 13 initiatives commissioned under an Innovation

Table 3.1 RIU Africa country programmes

Country	*Focus initiatives*
Nigeria	Cassava, cowpea/soybean, aquaculture
Tanzania	Dairy and post-harvest
Rwanda	Maize and potato
Zambia	Conservation farming and remote areas (isolation)
Malawi	Fisheries and cotton
Sierra Leone	Poultry and solar drying

Source: Internal RIU document

Challenge Fund working in Bangladesh, India and Nepal, with subsidiary activities in Cambodia and Vietnam. These projects would be clustered to reinforce thematic clusters across the countries around (a) dissemination of crop varieties developed through participatory research; (b) promoting use of current research in the value chain; and (c) promoting natural resource management research.

Global Best Bets

In line with the original thinking behind the RIU, a portfolio of Best Bets would be established. These would provide a significant level of funding (up to £1m each) to facilitate research outputs reaching their potential to provide significant benefit for the poor. Two of the proposed Best Bets had been pre-selected by way of the in-depth case studies undertaken by the programme over the previous two years into (a) rain-fed agriculture (Asia focused, originating from work under the RNRRS Plant Sciences Programme) and (b) that on the eradication of sleeping sickness (Africa focused, from work derived from the RNRRS Animal Health Programme). In the event the Best Bet programme was expanded to a larger number (see below).

I4D Fund

Privately managed small and medium-sized enterprises (SMEs) in agribusinesses currently play only a very limited role in taking research outputs to market in Africa. RIU took the view that the private sector needs to be encouraged, facilitated and in some instances created to participate in agricultural value chains. Strong SMEs can contribute to innovation platforms and form an important part of national innovation systems. In order to stimulate agribusinesses to begin using research results, RIU looked to establish an Innovation for Development (I4D) Fund.[4] A detailed scoping study for the I4D Fund was completed in early 2009, which established a market gap in the provision of early stage and investment capital for small and medium scale agribusinesses in Africa. However, this programme was discontinued in 2010 and replaced with an expanded Best Bet portfolio.

The lessons emerging from RIU's four intervention approaches would be collated to inform Output 2. The diversity of interventions was expected to generate useful evidence on how best to promote research, in order to achieve pro-poor impacts. An essential component under this output was to learn lessons on promoting research – what works, where, when and how. Research fellows were employed to undertake documentation and analysis of lessons in every case with intellectual oversight provided by two specified centres in Africa and India backstopped by a core team (Central Research Team [CRT]). The employment of research fellows was also intended as an explicit capacity building activity. The Research Fellows were expected to work closely with African country and Asian project teams.

Global Best Bet selection

The inspiration for the RIU Best Bets initiative came from the successful and popular BBC television programme *Dragons' Den*. Versions of this programme had been broadcast around the world under a variety of local names (*Money Tigers* in Japan; *Shark Tank* in the USA). The basic concept is that would-be entrepreneurs pitch their business ideas to a panel of wealthy and successful entrepreneurs who, subject to satisfactory due diligence, invest their own money and expertise in proposals that they find convincing in return for an equity stake in the business. RIU Best Bets took the central tenets of ideas being pitched to an expert panel, but in other significant aspects the procedure and principles varied significantly. A major difference was that the RIU Best Bets panellists would not invest their own resources; rather they make recommendations as to how RIU should invest its programme money.[5]

The objective of RIU Best Bets was to identify promising proposals to take existing agriculture research products and put these into use in ways that would benefit the poor (and others) in developing countries through partnerships in which private sector actors play a major role. The sum set aside for this in Africa was £5 million. Coverage would be on any aspect of agriculture in Africa – including crops, livestock, fisheries or forestry throughout the entire value chain, from production, through processing, storage and input and output markets, to consumption. The first round of RIU Best Bets was focused on East, Central and Southern Africa. A subsequent second round was held in early 2010 for West Africa. Lessons learned during the first round were incorporated into the West African round.

In September 2009 advertisements were placed in a number of newspapers covering East, Central and Southern Africa inviting the submission of Best Bets concept notes. Applicants were asked to limit these to two pages only; they would state how much financial support they were seeking from RIU, but no limits were specified. Concept notes were required to address four criteria:

- The proposal should be grounded in rigorous research in agriculture, including fisheries and forestry.
- The originators of the research should be involved in the programme in a significant way so that they would be able to apply their tacit knowledge and learning to the programme.
- The proposal was expected to achieve significant development impact at scale in East and/or Central Africa (and perhaps beyond).
- The proposal should comprise a consortium of partners (e.g. academic, public sector, NGO) led by an African institution and should include a private sector partner with evidence of support, which could be financial or in-kind.

In addition the following subsidiary criteria were also used:

1. Likely impact on incomes for poorer sections of populations
2. Likely impact on improving gender balance on these groups
3. Likely impact on employment creation
4. Ability of projects to foster the creation of networks that cut across professional hierarchies
5. Capacity of projects to foster local innovative capabilities
6. Capacity of projects to be sustainable after withdrawal of DFID aid.

By the deadline for submissions in early October 2009, RIU had received 105 concept notes. These were screened in a process in which RIU was assisted by the London-based Cambridge Economic Policy Associates (CEPA) - an economic and financial policy advisory business. A short-list of 11 proposals was developed. In two cases, pairs of proposals that appeared to offer significant opportunity for synergy (an army worm forecasting system and an army worm control technology; and two aquaculture proposals) were invited to amalgamate their proposals.

In another case, a proposal from Zambia to develop a value chain for cassava flour, in which the flour would be bought by a brewery and used as the raw material for a new type of beer, was put on hold. Although the proposal was highly rated by all the reviewers, guidance was sought from DFID because of potential conflict with its policy on supporting alcohol-related activities. The lead organisation for all the other short-listed proposals was asked to write a business plan following a format provided by RIU. To facilitate this, a grant of US$2,500 was made available which teams used in various ways, such as to bring team members together to enable them to work jointly on their plans. Two representatives from each proposal were also supported to attend the 'dragons den' event in Nairobi on 26 and 27 November 2009.

At this event, two representatives of each of the short-listed proposals presented their idea to the independent panel drawn from leaders in the African business, finance and research and development communities (see Note 5). The panellists had already read the business plans. Following a ten minute oral presentation, panellists had 20 minutes to interrogate the proposal, followed by a further 10 minutes in private to discuss the proposal among themselves. At the end of the day, the panel announced the proposals they were recommending that RIU should support. Subject to due diligence, which was then carried out during December by CEPA, RIU accepted these recommendations and proceeded to issuing contracts. Table 3.2 summarises the proposals that were considered at this meeting.

The money that RIU invested in the selected Best Bets was in the form of a grant since RIU's expected return on its investment was not financial; it was to be in the form of learning. The Best Bet proposals which RIU supported would thus become part of RIU's grand experiment in enabling innovation. RIU researchers would rigorously monitor the Best Bets with a view to teasing out useful lessons; what worked well, what worked less well and why? These lessons would then form an important part of RIU output and would help to shape future policy and practice to enable research to have greater impact on agricultural innovation. The Best Bet teams were also expected to work closely with RIU communication specialists and journalists to achieve widespread coverage of their research into use success stories.

At the end of this process it was then decided to extend selection to potential West African projects but because of time constraints selection was not done using a 'dragons den' approach. Instead, concept notes were solicited on exactly the same criteria and decisions were made internally by RIU management. In the event the west African proposals proved to be poorer than those from east Africa and ultimately only two were selected for support.

Table 3.2 Proposals presented to panel in Nairobi on 27 November 2009

Title	*Countries*	*Lead organisation*	*Other partners*
Village forecasts for combating Armyworm plagues: scaling up and out in East Africa	Ethiopia, Kenya & Tanzania	Desert Locust Control Organisation for East Africa	Animal/Plant Health Dept (MoA, Ethiopia); Plant Protection Services (MoA, Kenya and MoA, Tanzania); Juanco SPS and Bajuta Int. Ltd (both private), NRI, CABI Africa; Tropical Pesticides Research Institute
RIU safe and affordable Armyworm control for Africa (SAACA)	Tanzania	Ecoagricconsult Limited	Plant Protection Services (MoA, Tanzania), NRI and Lancaster University
Scaling-out tissue culture banana technology to enhance food security and livelihoods in Gatanga community	Kenya	African Harvest Biotechnology Foundation International	Technoserve, KARI, Aberdare Technologies Ltd (private) and Equity Bank
Marula and baobab food and beverages best bets for sustainable NTFP commercialisation in southern Africa	Zambia and Malawi	PhytoTrade Africa, PO Box BE 385, Belvedere, Harare, Zimbabwe	Afriplex Pty Ltd and Vital Solutions GmbH (both private)
Promoting yield improvement through farmer applied bio-control seed treatments in maize, sorghum and millet	Kenya	The Real IPM Company (K) Ltd, P O Box 4001, Madaraka, Thika -01002, Kenya	Univ Wales, Bangor, Uni Holenhiem and KARI
Increasing incomes of poor small-holder cassava farmers and processors through the development of high quality cassava flour value chains for the production of affordable cassava based beverages (**focus currently being discussed with DFID**)	Zambia	Zambian Breweries Plc (Subsidiary of SABMiller Plc); Plot 6438 Mungwi Road, P.O. Box 31293, Lusaka, Zambia. Tel: +260 211 244501	NRI

Title	*Countries*	*Lead organisation*	*Other partners*
Impact orientated public-private partnerships for reaching women and the poor with new bean varieties	Burundi, Tanzania and Uganda	International Centre for Tropical Agriculture, Kawanda Agricultural Research Inst. 13 Km Bombo Road P.O. Box 6247 Kampala Tel 256-414-567259	East & Central African Bean Research Network (ECABREN- CIAT); NARO, NASECO Uganda (private), Uyole ARI; Institute des Sciences, Agronomiques du Burundi and Ferme de Gihinga – Gatovu (private)
Aquaculture development: building services and supportive policy	Kenya and Uganda	FARM Africa	SON Fish Farm Ltd; Imani Development and KARI
Increasing fish production and enhancing food security through the dissemination of improved Tilapia strains and application of 'Best Practice' aquaculture techniques	Malawi, Kenya and Uganda	Bunda College	World Fish Centre, Project Concern International, Chambi Fisheries (Malawi); Kajjansi Aquaculture Research Development Centre; Walimi Fish Co-op Society (Uganda); Ministry Livestock & Fisheries (Kenya)
Transfer and dissemination of emerging agricultural technologies of new rice for Africa: improving access to seed through public-private partnerships in Uganda	Uganda	National Crops Resources Research Institute (NaCRRI); P.O. Box 7084, KAMPALA, UGANDA; Tel: +256-414-370907/ Mobile: +256 782 031285	NASECO (Uganda); Centre for Agricultural Inputs International and CABI Africa Matched funding available from Swiss Development Corp.
Empowering millions of smallholder farmers throughout East Africa to put research into use: a private sector led extension service to address climatic threats to food security	Kenya	Farm Input Promotions Africa Ltd. PO Box 5523, 00200-Nairobi, Kenya. Tel/fax: +254202730700; Cell: +254724700007;	Well Told Story Ltd, Minjingu Mines and Fertilizer Ltd; Athi River Mining Ltd; Leldet Ltd; Pannar Seed Co; Osho Chemical Industries Ltd; University of Exeter; University of Wales, Bangor

Selected Best Bets projects

The following Best Bets were selected for support in Africa:

i. Safe and affordable Armyworm control tools for poor farmers in East Africa to protect their crops against devastating Armyworm outbreaks
ii. Promoting yield improvement through farmer-applied bio control seed treatments in maize, sorghum and millet in Kenya
iii. Promoting the registration and use of Bio Control Agents (BCAs) in Ghana
iv. Aquaculture development in west Kenya: building services and supportive policy to increase fish production, and enhance food security through the dissemination of improved tilapia strains and application of 'best practice' aquaculture techniques
v. The promotion of minisett technology to improve clean seed yam production in Nigeria
vi. Transfer and dissemination of emerging agricultural technologies of New Rice for Africa (NERICA): improving access to quality seed through public-private partnerships in Uganda
vii. Empowering millions of small-holder farmers throughout East Africa to put research into use: a private sector-led extension service to address climatic threats to food security
viii. The use of multimedia to transfer technologies to young unemployed people through print media and radio in Kenya
ix. The use of new methods of livestock control to stamp out sleeping sickness (SOS) in Uganda.

All of these used RNRRS project outputs and all would be advised and backstopped by the original research groups. A summary of each project follows but full details can be seen in Chapters 4–6.

African Armyworm

The African Armyworm (*Spodoptera exempta*) is a moth which produces caterpillars in such high densities that they appear to march across the landscape. In fact African armyworm densities can exceed 1000/m^2 (Rose *et al.* 2000). The caterpillars feed on all types of grasses e.g. rice, wheat, millet, tef, sorghum, sugar cane and forage grasses. It is particularly active when rains have fallen after drought periods. The moths can migrate over hundreds of kilometres. Outbreaks occur almost every year in the primary outbreak areas of Tanzania, and under favourable conditions the Armyworm plagues can sweep across Tanzania and migrate to many parts of eastern and southern Africa. Although outbreaks are less frequent in Kenya and Ethiopia they still occur more than every other year. Armyworm is endemic in grassy areas, many of which are environmentally sensitive sites. This project's plan was to combine two technical solutions to a problem that has so far caused severe losses to poor farmers and has not been manageable by existing control and forecasting methods. The technologies are (a) *Forecasting traps* which would be purchased and disseminated to a number of villages in a number of areas (one to each village). These use pheromones to attract male moths and thus to indicate when an outbreak is due; and (b) *A Bio pesticide,* SpexNPV which would be produced by a private sector firm and then distributed to the villages where the outbreak is forecast. The technologies would be community based to achieve wide impact.

Real IPM: Stopstriga

Striga is a parasitic weed that grows into cereal crops, maize, sorghum, millet and rice preventing them from producing grain. Existing technologies have so far failed to deal adequately with a problem that has led to severe production losses for many poor farmers and as a result striga causes millions of pounds worth of damage in East Africa. The project leader is the Real IPM Company. Its plan was to promote the use of a biological control agent, a naturally occurring fungus which would be used to coat or dress farmers' seed and therefore give the plant protection from the striga parasite. This product had been trialled in a range of African countries and all responses had been positive. The seed dressing would be primarily marketed through farmer technology packs which included the mycoherbicide and instructions on 'farm seed priming'. This would be promoted through training networks, village promoters and mass media (vernacular radio), and would be made available from 'agrovet' shops. The project would utilise existing support networks as a mechanism for delivery of the extension message.

Ghana biological control

The RNRRS funded research which developed a number of biological control agents (BCAs) in Ghana: a pheromone for cocoa mirid, a pheromone for the cowpea pod borer *Maruca testulalis*, *Plutella xylostella* Granulosis Virus (PxGV) for controlling the caterpillar of the diamond-back moth on brassicas. A further RNRRS project, despite a short timeframe, had successfully worked with the Ghanaian Environmental Protection Agency (EPA) to develop registration protocols in 2005.[6] However, since the end of the project in 2006 until 2010 no products had been registered for use in Ghana. Consequently producers were still dependent upon chemical pesticides for pest control. Ghana's GDP benefits greatly from the export of cocoa and horticultural crops but to remain competitive in world markets they need to reduce pesticide use. There are 2,700,000 farm households in Ghana and over 80 per cent are involved in cocoa, tomatoes, cabbage, pineapple and mango production representing a significant market for pest control products. The Best Bet sought to catalyse the registration process in Ghana by registering four products, two viral and two fungal. The virus products would be produced by two east African firms.

Aquashops

The DFID RNRRS original research team had developed effective ways in India to encourage the private sector to drive the growth of the aquaculture sector and to create private-public partnerships in service provision called *aquashops*. This Best Bet was an amalgam of two separate proposals, one led by the University of Malawi and the other by the NGO *Farm Africa*. The overall rationale was straightforward. Since fish is an important ingredient of diet and since supply is weak in many African countries, a project designed to radically boost fish farming would play a significant role in poverty alleviation. Moreover the Kenya government had shown support through the construction of ponds in West Kenya. The model chosen was a franchise, whereby a core facility (the franchisor) would provide the technology and inputs needed to backstop six geographically located *aquashops* (the franchisees) who in turn would service the needs of the fish farmers themselves. In the event the NGO *Farm Africa* became the project leader and the project started in February 2010, rather later than many of the other Best Bets.

Clean seed yam

White yam (*Dioscorea rotundata*) tubers provide excellent nutritional and economic value as well as having a cultural importance for millions of people in West Africa and increasing numbers in East Africa. It is a high value crop with a large potential for increased productivity through increased commercial activity. Previous research funded through DFID CPP showed that pests and diseases are the major constraints to yam production. Because of these potential losses a key issue is the need for farmers to use clean (disease free) planting material but availability is a major constraint to yam production. As a result farmers often plant poor quality material already infected with pests and diseases which leads to low yields and infections spread to uninfected tubers. RNRRS research had developed the *minisett* approach which enables a multiplication ratio of 1:20 to 1:40 rather than the traditional range of 1:4 to 1:8. It had also shown that when farmers are provided with a loan or venture capital they are more likely to get involved in clean seed yam production via the minisett approach and for it to be a commercial success. The NGO Diocesan Development Services (DDS) in Kogi State began such a targeted scheme in 2002 with start up funds from Gorta (an NGO based in Ireland) and had supported 50 farmers. There was potential to scale-up this programme into other parts of Nigeria and indeed West Africa. The best bet would require ongoing technical support from IITA and NRCRI.

NERICA rice

NERICA rice was originally developed by scientists of the West Africa Rice Development Association (WARDA) and is a cross between an ancient, hardy African rice variety and a high-yielding Asian variety. It combines features of both: resistance to drought and pests, higher yields and more protein content than other types of rice. It has the potential to lead to reduced imports and increased income for farmers, thereby enhancing the well-being of rural families and reducing overall poverty throughout the region. However, the uptake of NERICA rice has not been as successful as desired. The Best Bet believed that the reason for this was that the various components of the production system (including the farmers themselves, their organisations and the private sector) were not sufficiently integrated. The team would be led by the Uganda National Crops Resources Research Institute (NACRRI) and assisted by two private sector seed companies and the international NGO CABI. It would use the RIU investment to increase NACRRI capacity for basic seed production, to increase capacity of targeted seed companies to produce multiplied output by establishing functional linkages with out-growers; and would stabilise product demand.

FIPS-Africa

FIPS is a not-for-profit company based in Nairobi, Kenya, largely donor-funded but also generating a small part of its own income.[7] Its mission is to improve the agricultural productivity and livelihoods of small-scale farmers in SSA. Here, marginal farmers produce very little on small farms (<1 ha), working on impoverished soils, using local varieties susceptible to disease and drought. Improved crop varieties are available at local research stations but farmers are generally unable to access them. Farmers often grow crops which are poorly suited to their environment, e.g. maize, which is often grown where there is not sufficient water or soil fertility. Poor farmers are unable to afford most farm inputs (in the standard retailed quantities), which in turn constrains the extent to which they are marketed in marginal areas. To address this issue FIPS-Africa had developed and implemented an innovative methodology to improve small farmers' access to the appropriate farm inputs, and information on best use. This has involved setting up decentralised small entrepreneurial advisory businesses that package and sell small quantities of clean inputs that reach as many farmers as quickly and cost-effectively as possible. To this end, it works closely with scientists in agricultural research centres and private sector seed and fertiliser companies. Close links with the private sector would ensure continued supply of farm inputs after funding has ceased.

Transfer of technology to unemployed youth: WTS-Shujaaz

Well Told Story Ltd. (WTS) was started in 2009 with support from the DFID (RIU) programme that provided 25 per cent of their income until June 2011. However, WTS is an entirely different instrument for getting research into use than the other Best Bets in so far as it uses multimedia to transfer technologies to people through print media and radio. Shujaaz was originally partnered by Farm Inputs Promotions Africa (FIPS), when the proposal was presented to the RIU in November 2009 as part of the Best Bets selection process. The plan was for the Shujaaz Youth Communications Initiative (SYCI) developed by WTS to transfer technologies using three forms of powerful and accessible youth focused media. These were (i) nationally distributed monthly comic books (300,000 every month); (ii) daily syndicated FM radio programmes, and (iii) interactive SMS to provide youth with ideas for generating income through agricultural technologies. After a short period, however, this Best Bet was allowed to operate on its own.

Stamp out Sleeping Sickness (SOS)

The Stamp out Sleeping Sickness (SOS) project had been operational for some three years and was included as a Best Bet because it satisfied all the selection criteria and had applied for funding to extend the project to a second phase. It focused on the use of pesticides and drugs to control the movement of disease-carrying cattle. And it was being run through a loose consortium based at the Faculty of Veterinary Medicine (FVM) at Makerere University, Uganda. Stakeholder interests include cattle owners, district veterinary officers, universities (mainly Edinburgh and Makerere), the corporate sector, local private sector companies, donors, and national regulatory authorities. The component being funded by DFID through RIU was an attempt to develop long-term sustainability through a public/private partnership designed to create small veterinary businesses, the argument being that unless steps are taken in this direction veterinary care will continue to remain within the province of charities such as NGOs and the corporate social responsibility (CSR) activities of international private companies. As such it will remain a 'dependent' activity reliant on outside inputs and in this sense not integrated into national capacity building. An important feature of this Best Bet was the link to student veterinary training at the University of Makerere.

Summary and conclusions

This chapter has provided a short historic outline of the RIU programme followed by more detail on its global Best Bet aspect. Inevitably, we have summarised what has been a complex activity taking place over a considerable time period. For those that wish to explore things in more detail reference should be made to the review and programme background documents given in Appendix 1 and the bibliography. In essence the Best Bet programme was an attempt to scale out technologies having potential value for poverty reduction using the private sector as a primary extension mechanism. An important characteristic has been its 'research/action' nature whereby 'learning' (output 2) has taken place in parallel with 'development' (output 1). Equally important, however, has been the use of public resources to leverage private resources outcomes which in turn produce sustainable development.

The RIU has thus been at least partially about a radical shift in technology development aid policy. Instead of funding scientific experiments and then leaving other bodies to put technologies 'into use' it has explicitly experimented with a much more integrated form of agricultural extension, one where research interacts directly with technology development. Since the earlier RIU programme was having problems in this regard a simplified RIU 2 was put into place. The various components of RIU 2 have all been experiments designed to learn lessons about how best to implement this type of technology development aid.

The remainder of this monograph will go more deeply into what we have learned through an examination of the funded projects. Chapters 4 to 6 deal with three projects each. Chapter 4 deals with Armyworm (Kenya and Tanzania), striga control/seed priming (Kenya) and BCAs (Bio Control Agents (Ghana)). All of them have been influenced by biological research and all have had a leading private sector dimension. Chapter 5 covers the Aquashops, Seed Yam and NERICA rice projects; these have been rather more NGO-led but still with an entrepreneurial component. Chapter 6 discusses the FIPS Africa, Shujaaz and SOS projects. Finally Chapter 7 will summarise the book's content and bring the monograph back to a policy perspective by drawing conclusions relating to the funding of technology development aid for very poor countries.

Notes

1 Again there is now an enormous literature on the idea of an innovation system as applied to agriculture in poor countries. A good summary account is contained in Hall (2006).
2 Internal RIU document.
3 Source: Internal RIU document.
4 The initial focus countries for the I4D Fund, as informed, by a feasibility study, were Tanzania, Uganda, Malawi, Rwanda and Senegal.

5 The RIU panel, who worked like the dragons on 'dragons den' were Judi Wakhungu, Executive Director of the African Centre for Technology Studies; Muchiri Wahome (Chair) Managing Director of Deacons (K) Limited, the leading chain store in the region; Patrick Oketa, Chief Investment Officer at the Kampala-based African Agricultural Capital and Ali A Mufuruki, Chairman and CEO of the Infotech Investment Group in Tanzania.
6 See R8430: http://www.cpp.uk.com/currentprojects.asp?step=9&pid=131
7 See www.fipsafrica.org/index.php for more information on this case.

4 Biotechnology projects

Introduction

This chapter summarises three of the projects funded by the DFID Research into Use programme (RIU) over the period 2009–2011. Each of them can be seen as an experiment in putting previously funded natural resource (RNRRS) research into productive use. Together they indicate how complex is the reality of capitalising productively on scientific research even where this has been commissioned for a useful purpose, in this case poverty alleviation in low-income countries (LICs). The projects deal with Armyworm (Kenya and Tanzania), striga control/seed priming (Kenya) and BCAs (Bio Control Agents (Ghana)). All of them have been influenced by biological research and all have a significant private sector dimension. The chapter then summarises project activities, the rationale for the interventions, the RNRRS projects that provided their scientific base, their co-operating organisations, objectives, progress and issues arising over the period. The final section provides summary conclusions and implications for policy.

Safe and affordable Armyworm control tools for poor farmers in East Africa to protect their crops against devastating Armyworm outbreaks

Area of intervention

This Best Bet was an amalgam of two separate proposals (led respectively by CABI-Africa and Wilfred Moshobozi, formerly of MoAFS Tanzania and now of Eco Agri Consultancy Services Ltd).[1] It has been implemented in Tanzania and Kenya. For Tanzania, agriculture is the foundation of its economy accounting for about half of its national income, three quarters of merchandise exports, is a major source of food and provides employment opportunities to about 80 per cent of Tanzanians. Agriculture in Tanzania is dominated by smallholder farmers cultivating an average farm size of between 0.9 hectares and 3.0 hectares. Food crop production dominates the agriculture economy; 5.1 million ha. are cultivated annually, of which 85 per cent is for food crops.[2] The major staples in Tanzania include: maize, sorghum, millet, rice and wheat, all of which are affected by African Armyworm *Spodoptera exempta* (Wilson *et al*., 2009). Since 1970, food production has expanded at a rate of only 2.9 per cent, while the population has grown at a rate of 3.3 per cent annually. However, much of this production gain has come through expanding areas under crops rather than improved productivity and, as unused land suitable for crops becomes scarcer, further gains from expansion are limited. During this period, Tanzania has changed from a net exporter of food to a net importer on a large scale. Similarly, Kenya's economy depends heavily on the agricultural sector. In Kenya, maize is the most important cereal although wheat, sorghum and millet are also important.[3]

Tanzania and Kenya are the countries worst affected by African Armyworm in Africa, growing over 5 million and 2 million ha. of cereals, of which 60 per cent and 75 per cent is maize, respectively. About 30 per cent of districts are susceptible to Armyworm outbreaks, or around 2.1 million households. The threat from African Armyworm is very real, as it is a particularly devastating pest that can devour any crop in the grass family, including pasture grasses, and so affecting livestock farmers. This is partly due to the Armyworms eating forage that would otherwise be available to livestock, but also because one of the main African grasses, star grass (*Cynodon*), is cyanogenic and releases cyanide in response to Armyworm damage, which has been known to kill young cattle (Georgiadas and McNaughton, 1988, Rose *et al*., 2000). Armyworm is a moth which can produce caterpillars in very high densities of 100 to 1,000 per square metre over large areas. They strip the land of growing crops and grass, as they appear to march across the landscape. They suddenly appear, catching farmers unawares and unprepared. This is due to the unusual migratory behaviour of the adult moths, which arrive en masse and each female can lay up to 1,000 eggs in her lifetime (Rose *et al*., 2000).

Uncontrolled outbreaks can cause total crop loss, with millions of hectares affected in bad years, although in some years there may be few outbreaks (Rose *et al*., 2000). Outbreaks occur almost every year in primary outbreak areas of Tanzania, and under 'favourable' conditions, the Armyworm plagues can proliferate and spread across Tanzania. Typically, after caterpillars devastate one area they pupate, changing into adult moths that later emerge to migrate to new areas in dense swarms which then lay eggs on the growing crops and pasture to start new outbreaks. This cycle of repeated outbreaks continues as the Armyworms migrate to many parts of eastern and southern Africa over 6–8 months of the year following the primary outbreaks in November. This coincides with the rainy season in which the main staple crops are grown. Although outbreaks are less frequent in Kenya and Ethiopia, they still occur more than every other year. Armyworm is endemic in grassy areas, many of which are environmentally sensitive sites. In May/June 2008 outbreaks of Armyworm were reported in 24 districts in Kenya – damaging 10,324 ha of crops and 41,435 ha of pasture. Between 2006 and 2009, Tanzania recorded outbreaks on crops covering 233,000 ha.[4]

The pest is seen as such a threat that current centralised forecasting systems and control are the responsibility of government. Although these are useful for national planning, they often do not help individual poor farmers. Farmers have also used imported synthetic pesticides to try and control outbreaks. Traditionally light traps, and more recently pheromone traps, have been used by government officers. When an Armyworm outbreak is forecast, pesticide or spraying services are requested from the appropriate government body. Unfortunately, even when there are resources to respond to the outbreak, the response is not always in time. The Best Bet was designed to supplement the centralised forecasting system[5] by introducing a forecasting methodology that would empower the poor. It would also replace the use of imported chemical pesticides which, due to high costs, limited access and environmental unacceptability, currently only meet the needs of around 30 per cent of poor farmers. In addition the forecasting traps would not only supplement the established forecasting system but would also give the farmers more time to prepare for attacks by utilising their own control or requesting control from government. The aim was to encourage the government to purchase the biological control agent at least initially.[6]

RNRRS projects

There were six relevant projects:

- *R6746 – Entomopathogenic viruses for control of African Armyworm Spodoptera exempta in Tanzania (The origin of the bio control product)*
- *R6762 – Decision tools to aid Armyworm surveillance and outbreak prediction*
- *R7966 – Identifying the factors causing outbreaks of Armyworm as part of improved monitoring and forecasting systems (The origin of the Community Based Armyworm Methodology)*
- *R8407 – Improved Armyworm monitoring and forecasting systems. (Also the origin of the Community Based Armyworm Methodology)*
- *R7954 – Novel technologies for control of African Armyworm on smallholder cereals in East Africa*
- *R8408 (2005) – Novel technologies for control of African Armyworm on smallholder cereals in East Africa.*

The technological history on Armyworm control goes back to 1965 (funded by ODA) and work had been ongoing during the 1990s. In addition to RNRRS work, 200 villages had introduced traps under an EU project involving Tanzania, Malawi and Zimbabwe.[7] The development of a strategy for local *SpexNPV* production itself drew heavily on lessons learned from earlier successful bio pesticide initiatives in India, where early involvement with the private sector was key to establishing commercial bio pesticides production (Grzywacz and Rabindra, 2001: 91–93). It also derived from analysis of the failure of previous bio-pesticides research in Africa to translate research into product development.[8]

RNRRS had produced two outputs on African Armyworm that were validated in the field and ready to be applied: a biological control product and a community based Armyworm forecasting methodology. The bio control agent, *SpexNPV*, is expected to be cheaper than chemical pesticides and safer for the environment. The adoption of *SpexNPV* will reduce the problem of stockpiling chemical pesticides. Tanzania, in common with many African countries, has a major issue in trying to dispose of thousands of tons of toxic chemical pesticides, originally procured or donated for migratory locust control, but which are now out of date and require expensive disposal to be made safe. It has been estimated that, Africa-wide, the cost of disposing of obsolete pesticides will be in the region of $1.2 billion.[9]

SpexNPV is a non-toxic, low-cost, locally produced bio-pesticide. It is a naturally occurring, highly infectious Armyworm virus that can rapidly wipe out outbreaks of the pest. Whilst *SpexNPV* is naturally occurring, it cannot be relied upon to arrive unaided/naturally, as it usually appears towards the end of an outbreak and towards the end of the Armyworm season, when most of the crop damage has already been done. However, it can be harvested and sold to farmers in affordable 50–100g bottles or equivalent small packs. This treatment is effective and is half the cost of chemical pesticides at around $5–6 per ha. Community based Armyworm control involves the use of forecasting traps which use pheromones to trap male moths. The discovery of a certain threshold number of moths over a period of 2–3 days indicates that an outbreak is imminent and gives the farmers time to take precautionary measures. The traps are issued one per village and operated by the communities themselves with the selection of representatives to manage the traps and methods of communication should an outbreak be forecast.

After the RNRRS work, outputs of the research on Armyworm were picked up by others. USAID supported community-based Armyworm forecasting (CBAF) in new sites in Tanzania and SADC, with EU support, funded a regional project with the regional body for Armyworm control (Desert Locust Control Organisation – DLCO), which continued until June 2010. These projects were focused on showing that CBAF worked in a greater number of physical locations and is a technology with a broad potential application. Similarly, USAID and DFID (with the BBSRC) have funded additional work on the ecology of *SpexNPV* as a potential bio-control agent for African Armyworm. This parallel research has helped to provide the evidence-based wide-scale strategic control of Armyworms in eastern Africa. These results were shared with DLCO and national staff. The use of *SpexNPV* was given impetus by the 2004 agreement from Tanzania's National Plant Advisory Committee to adopt *SpexNPV* as a replacement for chemical pesticides for Armyworm control.

The Armyworm Best Bet has been assisted considerably by continuing UK Research Council (BBSRC) support for Lancaster University (Prof K Wilson) and NRI (Dr D Grzywacz). For example, Dr Grzywacz helped draft the production and applications manual for the Arusha facility and Prof Wilson has developed a database to help target the high risk areas. This database collates more than 25 years of information from Armyworm outbreaks and from moth traps throughout Tanzania. These data are available in a number of user-friendly tables, graphs and maps, and can be used by the MoAFS national Armyworm forecaster and staff, as well as by Eco Agri Consult Ltd., to help identify areas at greatest risk of Armyworm attack.

The project

The following bodies have been involved in this project:

- CABI Africa, Kenya (CABI)
- Eco Agri Consultancy Services Ltd, Tanzania (EAC)
- Ministry of Agriculture and Food Security, Tanzania (MAFSC)
- Ministry of Agriculture, Kenya (MoA)
- Natural Resources Institute, UK (NRI)
- Lancaster University, UK (LU)
- Desert Locust Control Organisation for Eastern Africa, Ethiopia (DLCO-EA)
- Pest Control Products Board, Kenya (PCPB)
- Tropical Pesticides Research Institute, Tanzania (TPRI)
- Bajuta International, Tanzania (Bajuta)
- Juanco SPS, Kenya (Juanco)
- Ministry of Agriculture and Rural Development, Ethiopia (MoARD).

The overall objective has been to establish a system for the production, supply, distribution and marketing of Safe and Affordable Armyworm Control Tools (SAACO–Tools). The tools are for local forecasting of outbreaks, and a cheap, safe biological pesticide for controlling them. Together the tools will reduce the devastating effects of Armyworm outbreaks on food production. In sum the objectives were to:

i. Establish a supply network for registered, low cost forecasting tools, meeting the full needs of Tanzania and Kenya (estimated at over 5000 villages).
ii. Establish a virus production system in Tanzania capable of producing at least 10,000 ha worth of product per annum, and able to expand to meet the regional need of >100,000 ha per annum.
iii. Market SAACO-Tools to customers, including government services, farmers, community organisations, NGOs and development partners, for sustained use and expansion to all affected countries in Eastern and Southern Africa. This would include establishing local forecasting in 120 villages, and applying the virus on up to 1000 ha.

Progress

Techno-economic

A private sector *SpexNPV* production facility has now been established in Tanzania with pheromone sourcing from a UK company (Russell IPM – which has now opened an office in Nairobi).[10] The biotechnology facilities in Tanzania are unique and well adapted for production of other biotechnology products needed by the Tanzanian export agriculture sector such as plant tissue culture for bananas, pineapples and potatoes for which customers (including a USAID Tanzania Agriculture productivity project) have been identified. These developments will provide a small number of jobs directly in production, but the production of new pest control and biotechnology products will have greater impact through strengthening the export sector of Tanzanian horticulture by providing inputs (bio pesticides and tissue culture stocks) needed to strengthen export competitiveness which Tanzania currently lacks. Expectations of beneficiaries are outlined in Table 4.1.

In addition there is also a request from a private large-scale business, Kilombera Plantations Ltd., who need Armyworm forecasting services and *SpexNPV* control for 5,000 ha of rice, which has in recent years been heavily attacked by African Armyworm. Primarily the project will target outbreaks in central Tanzania reaching about 500 small-scale farmers (under contract) and two big commercial rice farms. Community based Armyworm forecasting enables farmers to make forecasts for their own village with over 80 per cent accuracy. The trap and chemical attractants are expected to cost some $14 a year. One forecasting tool can cover around 200 ha and has an effective life of around 2 months. It is therefore assumed that this project will be profitable as the value of crop loss is likely to be considerably greater than cost of remedial technologies.[11]

The establishment of the *SpexNPV* facility is one of the first commercial biological pesticide plants in Tanzania (only the third in Sub-Saharan Africa outside RSA). It provides the only source of environmentally safe, affordable

Armyworm control in Africa; it is also a platform for developing new businesses in safe pest control products and active discussions are underway already to enlarge the range of products to be produced to support improved pest control both by poor farmers and export agribusinesses in Tanzania.[12] At government level Kenya has now approved a simple (no cost and fast) procedure for the registration of pheromones and the bio pesticides. This will facilitate the commercialisation of other pheromones for lepidopteran pests in Kenya. A partial forecast tool supply chain has now been established, with the distributor and producers signed up.

There has been a draft inclusion of community based Armyworm forecasting on the Government of Kenya (GoK) permanent secretary's (agriculture) contract, meaning that getting CBAF adopted is a key objective for him. Already the impact of this is being seen with mention of government funds being made available for CBAF. There has been a change in mindset with the Government of Kenya now recognising that CBAF has an important role to play in providing data on Armyworm status to add to that from the national network of traps. This will enable more accurate forecasts to be made and control activities better pinpointed.

Table 4.1 Expected beneficiaries

Number & Type of Indirect Beneficiaries	*Number & Type of Direct Beneficiaries*	*Male Beneficiaries (indirect and direct)*	*Female Beneficiaries (indirect and direct)*
	80,000 in Kenya benefitting through forecasting	Approx 40,000	Approx 40,000
	25,000 farmers in Tanzania benefit through forecasting under RIU. The number is bigger if previous initiatives by other projects are considered	12,500	12,500
The full impact of *SpexNPV* facility will be felt after *SpexNPV* is produced and distributed to farmers in 2012			

Source: Internal RIU Document

In Tanzania, government support has been both financial and in terms of adding additional sites for CBAF. CBAF has been integrated into district agricultural development plans (DADPs), thus ensuring longer term sustainability. The Government of Tanzania (GoT) has also committed to using *SpexNPV* in its Armyworm control programmes to show farmers its efficacy. Gildemacher and Mur (2012), however, remain uncertain about the capacity of Tanzania's national extension system to contribute to this objective.

Capacity

On capacity building, harvesting techniques were validated in the trials and improved methods for mass harvesting developed.[13] Training of trainers' courses took place with 112 trained in Kenya and 40 in Tanzania. CBAF was established in 120 villages (80 in Kenya and 40 in Tanzania). A CBAF conference was held in Kenya which increased awareness. There was press coverage on various issues and collaboration was initiated with Well Told Story Ltd. (see Chapter 6) to include issues on CBAF in their papers; there have also been talks with FIPS Ltd. (see Chapter 6) about training their staff as forecasters. Finally there have been films, conferences and articles written about Armyworm and the Armyworm project. A website has been created that gives both background and up-to-date information on Armyworm news, including a re-posting of the latest Armyworm outbreak forecast for the region from ETOP.

Partnerships/linkages

The following partnerships/linkages have been established.

- The team has worked closely with relevant policy makers and have arranged meetings with Ethiopia and DLCO-EA who are already involved as regional partners and are publicising CBAF. There are discussions with the governments to increase the purchase of pheromones for wider usage. The Malawi Ministry of Agriculture has expressed interest in establishing a distribution system for *SpexNPV* once the product is ready. The Tanzanian facility has also been approached by BMGF with a view to assisting in cassava and sweet potato protection.
- There may be difficulties in developing effective Armyworm control measures if reliance is placed solely on government support for the market system since extension services in both Tanzania and Kenya are weak. Most interviewees agreed that greater reliance may need to be placed on initial private sector led investment, supported in places by donor funding. However, the Kenyan Ministry of Agriculture,

especially the Plant Protection Division (PPD), has now bought into forecasting and is trying to help as much as possible. It is they who have set up the 'alert' system whereby farmers can communicate outbreaks directly to the PPD who then inform extension agencies to spread the word. There is no indication yet of how successful this measure is.

- There has been an approach from a UK venture capital firm interested in backing the private sector facility in Arusha.

Promoting yield improvement through farmer-applied bio control seed treatments in maize, sorghum and millet

Area of intervention

Striga is a parasitic weed that causes a significant loss of yield in maize, sorghum and millet in Western Kenya. Referred to as 'witch weed' due to its devastating effects on crops, it attaches itself to plant roots from which it draws its moisture and nutrients, inhibits plant growth, reduces yields and in extreme cases, causes plant death. More than 80,000 ha of maize are reported to be affected in Kenya alone, and yields can be reduced by up to 80 per cent. Striga infestation is most severe in Nyanza and Western Provinces, where it occurs in about 72,800 ha and results in crop losses estimated between KSh 800,000 and KSh 2,200 million per year (Manyong *et al.*, 2007). The impact of striga is greater where soils are less fertile, and so in areas of soil denudation it is an increasing problem. Farmers respond to striga infestation by hand weeding and, less often, burning affected fields but the efficacy of these practices remains questionable; this is so given the large numbers of seeds that a single, mature plant produces and returns to the soil, and the fact that these seeds can remain dormant in the soil for up to 15 years.[14]

Other methods for reducing the impact of striga have been developed, for example improving soil fertility, crop rotation of infested land with non-susceptible crops or fallowing. Rotation with non-host crops interrupts further production of the striga seed but striga can remain viable in the soil for many years (Riches *et al.*, 2001). However, this approach has come under increasing demographic pressure with consequent increased demand for food; so there has been intensification of land use, mono cropping and decline in soil fertility.

More recently use has been made of seed resistance to systemic herbicides and suicidal germination induced by non-host plants, alongside complementary packages such as crop rotation; this has been more successful. These technologies have not been widely adopted due to farmers' socio-economic conditions. In addition poor adoption is due to the inadequacy of any one single component for sustainable striga control and the absence of integrated management strategies suitable for existing farming systems. There is need therefore to adopt a farming systems approach for the development and implementation of integrated striga management strategies suitable for the various agro ecosystems. It has been recommended that efforts should be directed towards identifying herbicides that are cheap, persist in the soil and are compatible with the mixed cropping systems practised by most subsistence farmers.

RNRRS projects

There were four relevant projects:

- *R6395 – The development and testing of seed-priming to improve stand establishment, early growth and yield in semi-arid Zimbabwe and India (1995–99)*
- *R7189 – Cultivar competitiveness and interactions with on-farm seed priming for integrated weed management*
- *R7440 – The physiological basis for the effects of on-farm seed priming in tropical crops: interactions with seedbed physical conditions*
- *R7438 – Participatory promotion of on-farm seed priming.*

This Best Bet made use of previous RNRRS research that aimed to provide crop and cultivar-specific parameters for seed soaking of sorghum, pearl millet, maize, upland rice and cotton. It had tested the practicability of the technique in on-station trials and in farmers' fields in Zimbabwe and India (R6395). As a result of this study on-farm seed priming has been shown to be a viable key technology – low cost, low risk with immediate benefits that can persuade farmers to adopt further improvements. It was also found that participatory research has been essential for the successful promotion of the technology.

Further research aimed to test the crop physiology underpinning 'on-farm seed priming'. It was established that previously reported benefits of priming in the field could be demonstrated in controlled environments for maize, rice and wheat (R7189). In experiments in the laboratory, controlled environments and in the field, there was no evidence that priming inherently changed the way that plants grow. Instead, the benefits depend on the extent to which priming can advance germination and emergence. This in turn depends on the soil physical conditions at sowing. In participatory trials with farmers, the faster emergence from primed seed was recognised and appreciated as primed crops were considered to be more competitive against weeds. This output provided assurance that priming can improve yields through increased crop stand, more rapid seedling emergence and increased uniformity of the stand.

Another study (R7440) aimed to investigate key issues likely to shape the future direction of crop protection practices, such as impact and sustainability of new technologies in crop protection; and approaches to improve the effectiveness of farmer crop protection practices were identified, with appropriate guidelines developed. Researchers undertaking these experiments in different institutions concluded that there is a need to first understand the mechanisms involved in interactions between priming and rice/weed competition, before any general recommendations about the use of priming in weed management can be made.

The RNRRS researchers also looked at ways of promoting participatory on-farm seed priming, by identifying various pathways, piloting them and promoting the most successful (R7438). The researchers were looking to see what could be done to increase the proportion of seeds that emerge, and the rate at which they do so, which will have a large impact on farmer livelihoods. Low, unstable yields are a major contributor to the fragile lives of poor farmers in marginal areas. Low-cost, low-risk interventions that increase and stabilise yields will have a large impact on the livelihoods of such farmers. Finally research from University of Hohenheim and IITA isolated the fungus that was the basis of the proposed bio control; it undertook its characterisation and undertook limited field testing in West Africa. In addition IITA identified other strains of the fungus and worked on these as a comparison.

The project

The project has been led by the Real IPM Company (K) Ltd., which was established in 2004 and is based on an old coffee farm near Thika, Kenya. Real IPM's income comes from training, consultancy and biological control agents (BCAs). Almost 80 per cent of its income comes from the sale of BCAs to Kenya's export horticulture industry. The company has two categories of BCAs (i) predatory mites; and (ii) beneficial microbes (bacteria or fungi). It breeds and produces 10 to 20 million predatory mites per week.[15] It had a turnover of Ksh 71 million in 2009 and now employs over 80 permanent staff. In Kenya about 50 per cent of flowers are treated with bio pesticides and about 50 per cent of these are supplied by Real IPM. Ten years ago bio pesticides were hardly used but are now increasingly in demand due to retailer standards and international regulatory pressures. The 2008 recession spurred Real IPM to consider alternative markets to export horticulture and identified the millions of smallholder farmers in Kenya as an alternative market.

However, without donor support, Real IPM was unwilling to enter this little understood market (there are no BCAs registered for use on domestic cereal crops in Kenya). The registration procedure was supported by the DFID Agriculture Enterprise Challenge Fund (AECF) which provided a grant and a refundable loan. Real IPM has undertaken project management and coordination of the six activities and has employed an MSc graduate, who is based in Western Kenya and speaks the local language, to manage and coordinate the project in the field. Other partners are:

- Bangor University and the University of Hohenheim are key partners involved as originators of the research and provide technical backstopping. Moreover the University of Hohenheim has wide experiences in GAP for striga management and in striga training. They have been involved in development, production and promotion of small farmer technology packages.
- Greendown House Ltd is a company specialising in media and market communication. They have led the activities involved in communication.
- KARI (Kibos) has been undertaking research in maize germplasm evaluation, maize agronomy and striga control research and were partners in other RNRRS and striga push pull/GAP projects.[16]
- Real IPM has informally approached other network providers including MoA and NGOs working in the sector like FIPS, Sacred Africa and Farm Concern. All have expressed an interest in participating in the network of support for the farmers to extend the technology packages.
- Radio Lake Victoria and Ramogi radio, the latter being part of Royal Media Services Limited, Nairobi, have collaborated in the project and are familiar with operating SMS platforms in conjunction with radio programs.

The project objective has been to promote the use of a technology package to small-scale farmers who save their own cereal (e.g. maize) seed. The two major components are the use of a mycoherbicide to control the parasitic weed striga, and the use of seed priming. More specifically, project objectives include:

- The establishment of a support network and training in the technology for 100 field workers of national programs, NGOs, etc.
- Promoting the technology through radio programs and advertising, and communication via an SMS messaging service. Establishing an SMS database of up to 48,000 participants. Additional promotional activities – demonstration plots, leaflets, posters and stands at agricultural shows.
- Support for farmers in three successive waves of plantings in Nyanza province by distribution of small scale technology packs to 48,000 farmer households.

The chosen region was Nyanza province, which had a population in 1999 of 4.4 million comprising 968,014 households. Four districts (half Nyanza's total) were targeted. Many of these households are cereal producers; hence the total target market was about 480,000 households.

Assuming a 10 per cent technology uptake, 48,000 households were using the new and innovative technology by April 2011. For maize, the mean financial benefits of this project to small-scale farmers have been estimated at Ksh 1,800 per hectare or 250 kgs/ha of increased maize yield per crop with more in heavily striga infested fields. The estimated value to the rural economy of Nyanza was some Ksh 25.9 million in one planting season assuming an uptake of 10 per cent of the households in the target area.

Progress

Techno-economic

Stopstriga pot trials were completed and a report submitted to the PCPB. Field trials demonstrated that Stopstriga does not inhibit or slow maize seed emergence. A commercial source of phosphate for seed priming was identified which adheres to cereal seed and is effective in boosting early germination from seed priming. This led to the development of a product *Gro-Plus*. Results showed that Gro-Plus has a positive effect on emergence and plant development. Data on height and number of leaves were taken 14 days after sowing. At the same time the cost of production was seen to be too high for the small farmer but further work was done to modify the product and reduce its price. The product was refined and improved with better formulation and a better sticking agent (glue). Trial plots showed much better results. There followed the submission of a dossier of information to the Kenya Bureau of Standards for Gro-Plus to obtain approval for sale and a quality stamp.

Over 50,000 farmers in Western Kenya were recruited and registered to conduct trials of Gro-Plus; these collected feedback and helped with the marketing of the product. The details of 48,000 farmers trained including their mobile telephone numbers, have been recorded on the project website[17] as part of a database that will be used to communicate with farmers using SMS messaging and phone calls. Real IPM report that 45,000 farmers have received Gro-Plus packs[18] at a retail value of Ksh 130 each and 54 per cent of the farmers reported improved growth and yield based on the telephone survey. There are as yet no data to say income has increased but this now appears plausible.

The project has evaluated the use of local vernacular radio and the provincial administration (chiefs' barazas) to recruit farmers and promote the technology, with the latter being more successful than the former. Other methods have included advertising and SMS messages. Many organisations have been surprised at the number of farmers recruited through the Provincial Administration and the chief's 'barazas' to the SMS data base. This is a clear lesson and mechanism that could be exploited for other projects trying to access small-holder farmers. The approach might not be effective in all provinces but it is currently being evaluated in different provinces in Kenya. For distribution, the team have recruited 46 Agrovets who are buying anything from 1 to 200 packs of Gro-Plus to sell on to farmers.

The PCPB has restricted registration trials of Stopstriga to the greenhouse in the first instance. This Best Bet has shown that there is a real issue with getting BCAs registered for use in Kenya. Although there are guidelines, the regulatory bodies seem not to follow either a pragmatic or an efficient approach and communication between KEPHIS and PCPB could be improved. The registration of Stopstriga has thus been delayed causing a greater focus on Gro-Plus seed priming. At the same time the delivery system for Stopstriga has had to be revised since it is not offering effective control (much poorer than the results from the field in Ghana [see below] where the fusarium isolate originated). Real IPM is now looking for a local isolate from striga infested fields in Western Kenya. The value of selecting a Kenyan isolate of Stopstriga (*Fusarium oxysporum*) is that it will not require regulatory permission to do field research.

There is some evidence that the use of Gro-Plus, even without the myco herbicide, actually helps to reduce striga infestation (from the University of Wageningen and comments made by farmers). Real IPM is now exploring this with a small group of 100 farmers looking specifically at the effect of Gro-Plus on striga infestation. Real IPM is now considering the application of Gro-Plus to other crops including vegetables and also to be used in machine planting by larger cereal farmers.

There are important lessons to be learnt in terms of marketing farm inputs to farmers who previously did not buy them. For example, the distribution of inputs to isolated small-scale farmers is a challenge and the limitations of the 'agrovet' system are being discovered. Agrovets are not evenly distributed; farmers often have to travel 15–20 km to the nearest agrovet. Also they can be part-time, have limited stocks and lack product knowledge; therefore they are not always a reliable distribution system. As a result Real IPM is now considering linking with a larger agrovet company (Farmchem) for distribution purposes, though there are obvious intellectual property dangers here. Another approach is to train identified agrovets on how to prime seeds. This is being done experimentally with around 400 agrovets. Finally, arrangements are now in place with another RIU Best Bet (FIPS Ltd) to sell Gro-Plus at a reduced price to some of its market agents for use in the field. In fact, promotion and dissemination through meetings organised by SMS messaging seem more effective than depending on the agrovet system to provide inputs. Interaction with farmers has led to changes – looking at easier ways of delivering Stopstriga and fertiliser in a useable and easy way to the farmer. It has been exploring methodologies, delivery, costs and farmer acceptability.

Capacity

On capacity there has been the establishment of a network of support and training for farmers in Western Kenya. To enable the programme to be implemented, a series of meetings were held and agreements obtained from various government departments (e.g. Provincial Commissioners, District Agricultural officers). Meetings were held with various local NGOs, institutions and self-help groups and a data base was collected of existing networks in the area. The project also identified and used village promoters who are the conduit for dissemination of information, and inward feedback in building a database of the mobile phone contacts of interested farmers and smallholders.

Linkages/partnerships

There are important linkages with many other Best Bets (especially FIPS) and of course the wide range of groups cited in the following project that is also managed by Real IPM.

Bio control agents (BCAs) – Ghana

Area of intervention

This Best Bet on bio-pesticide control in Ghana has been led by the Real IPM Company and builds on the other Real IPM-run project supported by RIU in Kenya (discussed above). It focuses on reducing crop losses from pests which in turn should lead to improved food security, income generation and improved livelihoods. In addition, consumers and the environment should also benefit from reduced use of pesticides, and hence lower pesticide residues in food, reduced contact with pesticide and impact on non target organisms. Although systems for registering chemical controls for efficacy, health and safety standards are generally in place, they are often time consuming and costly, affordable only by large pharmaceutical companies. Very few deal with the specific requirements of registering biological control agents for which simpler, usually lower cost requirements could be applied. Kenya is one country which does have such protocols, which were developed through the RNRRS and implemented with Government of Kenya support, with particular consideration for the export horticulture industry (Wabule *et al.*, 2004). Retail standards (e.g. GLOBALGAP [previously EUREPGAP]) and legislation (e.g. EU's Maximum Residue Level) to protect the consumer from excessive pesticides are now implemented in most developed markets. Compliance with these rules and standards is critical for those wishing to export to these markets.

Currently, in Kenya the horticultural industry is the fastest growing agricultural sub sector in the country, and contributes 23 per cent of GDP, and this growth was originally based on sourcing product from smallholder farmers.[19] Across Africa there is potential for growth and poverty alleviation through the expansion of the fresh produce sector. A fundamental component of this type of voluntary standard is the requirement for producers to adopt integrated pest management (IPM) and so reduce pesticide inputs in response to market demands. In Ghana, it is estimated there are 2,700,000 farm households, and over 80 per cent are involved in cocoa, tomato, cabbage, pineapple and mango production. For them to be able to exploit export markets they will need to insure low levels of pesticide use. This is possible through Integrated Pest Management (IPM), the adoption of which has been mixed in Africa. Increasingly, the potential of biological control agents (BCAs) for use instead of synthetic pesticides (but without the harmful side effects of pesticides) is recognised.

Demand has also been articulated by the public sector; in Ghana the Environmental Protection Agency (EPA) itself was anxious to be ready for anticipated demand for biological products, and given the importance of IPM as a national strategy for crop protection, this demand is reflected in national strategy. Indeed it requested, and was provided with, support from the RNRRS (Cherry, 2005), however until the RIU, no product had been registered using these protocols. The RIU, by funding this Best Bet, recognised the possible dichotomy of having a number of products developed via research projects; yet an inadequate regulatory framework to deal with them, effectively placing a barrier to production. One of the keys to successful wider implementation of specific regulations will be the negotiation of common or harmonised standards of registration, or equivalence between dossiers. This would not have to preclude national sovereignty on regulation, but it would favour product development and commercialisation. Because of the fragmented nature of African markets and the number of African countries, steps that facilitate the registration process on a regional scale should make product development more economically feasible.

RNRRS projects

The DFID Crop Protection Programme (and also its Crop Post Harvest Programme) had invested in a number of projects throughout the RNRRS that developed or transferred biologically based crop protection technologies for developing countries. In some cases there was concern that, given the absence of appropriate regulatory frameworks for registration of the technologies generated by RNRRS projects, adoption of these technologies would be hindered. Indeed, it became widely accepted that one of the major barriers to the implementation of biologically based pest control technologies was the absence of registration guidelines. This was certainly the case for project R7960 in which national stakeholders in Benin and Ghana concluded that the most important next step to the implementation of a viral bio pesticide against the diamond back or cabbage moth (*Plutella xylostella*) for W. Africa brassica production would be to establish specific bio pesticide registration guidelines, since at that time regulatory authorities only had frameworks for registration of synthetic chemical pesticides. In Ghana, the process of developing a bio pesticide registration framework began in 2004.

In more detail the relevant RNRRS projects were as follows:

- *R7249 – Development of myco-insecticides and pheromones for cocoa mirids*
- *R7441 – Development of pheromone trapping – monitoring and control of the legume pod borer by farmers in West Africa*
- *R7449 – Developing an improved sustainable integrated pest management system for brassica production in Kenya*
- *R7960 – Public-private partnerships to develop viruses as bio-insecticides for Ghana and Benin*
- *R8300 – Implementing pheromone traps and other new technologies for control of cowpea insect pests in West Africa*
- *R8313 – Integrated pest management for Cocoa in West Africa*
- *R8430 – Bio pesticide registration and risk assessment guidelines for the Ghanaian EPA.*

Project

This Best Bet included some of the original researchers from the RNRRS based at the Natural Resources Institute (NRI (Chatham)); so it has been able to access both tacit and codified knowledge required for successful implementation. In particular, these researchers have been involved with facilitating the necessary legislation in Ghana and have worked extensively on baculoviruses for control of *H. armigera* (cotton bollworm) and *P. xylostella* and BV registration for the DFID Crop Protection Programme. They provide technical support for registration authorities in evaluating registration dossiers and with international standards in export horticultural crops. The main project partners are:

- *CABI UK* has developed IPM systems in cocoa; they are key partners involved as originators of the research and provide technical backstopping. CABI have considerable experience in developing BCAs in tropical situations and crops, and have worked extensively in West Africa.
- The *Real IPM Company Ltd* and *Kenya Biologics Ltd* are BCA production companies in Kenya and have commercial experience of registration. They produce the BCAs in Kenya, prepare the dossiers, assist in trial protocols and provide trial product samples.
- *Environmental Protection Agency* (EPA) is the authority in Ghana for the registration process which includes evaluating the dossier

information, supervising the efficacy trials and presenting the information to the Pesticide Technical committee.
- *Pest Control Products Board (PCPB),* the Kenyan regulatory authority shares 'best practice' with the Ghanaian authority.

Following a visit to Ghana by Real IPM as part of the activities related to preparing the dossier, the following additional partners were identified:

- *Wienco (Ghana) Ltd* (www.wienco.com) is an agro-chemical company very keen to collaborate and act as a distributor company; they have good relations with the Cocoa Board/CRIG having already registered agrochemicals using these as testing agents. Recently they launched their Cocoa Abrabo-pa package (Abrabo-pa = a better life).
- *Agro Eco Louis Bolk Institute* (www.louisbolk.org) an organisation with an office based in Ghana that sponsors and promotes organic produce in particular organic cocoa, which is exported to the European market, again keen to collaborate.
- *Blue Skies Products (Ghana) Ltd*, (www.blueskiesproducts.co.uk) a fresh produce company that prepares and exports fresh cut fruit salads using mango, banana, pineapple, passion fruit and papaya. Due to reduced availability of pesticides they were keen to employ BCAs.
- *Mr Emmanuel Owuso,* a private consultant who has worked with the EU's pesticide initiative programme (PIP) (www.coleacp.org/pip) and has been trained by Real IPM in the past. He would act as coordinator in Ghana for the project.

The overall aim of this project was to register (official authorised use) and promote the use of four BCAs in Ghana: a) Two baculoviruses for control of bollworm (*Helicoverpa armigera*) and DBM (*Plutella xylostella*); b) a fungus (*Trichoderma asperellum*) for the control of *Phytophthora spp.* in pineapple and black pod disease in cocoa; and c) a fungus (*Beauveria bassiana*) for the control of mealy bug in cocoa and fruit crops. These products are to be sold as income generating products by the producers (Kenya Biologics and Real IPM). They will be distributed on commission by Ghanaian companies that provide inputs supplies, and adopted by farmers as they are recognised as offering value for money. In addition to these particular BCAs the project was expected to open the door through promotion for other bio rational technologies to be registered and used in Ghana. In summary the main expected objectives have been:

1. the registration (full or provisional) of four BCA organisms in Ghana
2. the promotion of the registration process by holding five workshops and publishing a booklet on the registration procedure for BCAs
3. the development of a distribution/marketing network for BCAs with a signed distributor agreement.

Progress

Techno-economic

All four BCA product dossiers were submitted to Ghana's EPA for review. Permission was given for the University of Ghana to conduct efficacy trials, and samples of three BCAs (*Trichoderma, Metarhizium and Helicoverpa baculovirus)* were safely, and with full permission of the authorities (Kenyan and Ghanaian), delivered to Ghana. Also, samples of baculovirus were sent to a GPL laboratory in India for eco and toxicology testing. This resulted in *all four of the BCAs being successfully registered*, and there has been improved clarity of forms and awareness raising activities. With registration achieved, attention turned to the development of a distribution network for BCAs with a distributor identified, and marketing commenced for some BCAs at the end of May 2011. Visits to Ghana identified potential markets and actual distributors for the products, which will enable the products to be marketed.

The main thrust of the work has been in working with those implementing the policy (EPA) previously developed under the RNRRS, testing how appropriate it is and giving the regulatory authority experience in the registration process. The team have also reassured the regulatory authorities in Kenya (KEPHIS – the Kenyan Plant Health Inspection Service) that exporting Kenyan isolates is not a threat to Kenya. As a result, a good working relationship has been developed with the registration authorities in Ghana which is leading to swift processing of effective efficacy and toxicological procedures, leading in turn to a fast registration process. The registration is far faster than in Kenya; potentially it is a model that could lead to fast and efficient registration in other African countries. The project has resulted in the EPA and the Ghanaian registration committee working – for the first time – with the private sector in the registration of BCAs in Ghana. Two Kenyan companies have forged relationships with the regulators, distributors and potential customers in Ghana forming new South–South linkages. The effect has been positive and it is hoped will lead to expansion of this market in the future. The University of Ghana, Crop Science Department who undertook the efficacy testing, have also reported they have learnt from the exercise of working with bio pesticides.

The local distributor (Wienco) was selected for the two Real IPM fungi, and opportunities for entering the more lucrative cocoa and oil palm markets are now being explored. Kenya Biologics and Real IPM have each visited Ghana and held discussions with potential distributors, Wienco, Dizengoff, Calli Ghana and RMG in Cote d'Ivoire. Real IPM decided to partner with Wienco (as a pesticide distributor) because this company knows how to get crop protection products into cocoa in Ghana, and has already achieved this. However Real IPM believes that registering products for this sector will take longer than the duration of the RIU funding. They have also partnered with Agropharm West Africa Ltd, as another distributor. The team has also been in discussion with a number of other companies interested in using their products (e.g. Unilever) but as yet a way forward has not been identified. Some further points of interest are as follows:

- A problem encountered was whether cold storage is available or not to retailers in Ghana and this could affect whether or not the virus BCAs can be marketed. Market size is a key issue because the distributors are keener on the products that Real IPM are offering that have a larger market opportunity (by crop [including its market value] and range of pests) than the more pest specific products Kenya Biologics are offering (more suitable to smallholders).
- Real IPM is investigating the possibility of marketing the BCA 'spores' to Ghana as a means of reducing the price. Subsequent formulation would then be done by the distributor.
- One of the baculoviruses (against the diamond-backed moth) has not proved effective.
- The *Metarhizium* fungus for treating fruit trees is being sold by Real IPM in liquid form under the brand name 'Campaign'. It may also be effective against the alien fruit fly *Bactrocerus invadens*, which is now threatening much of Africa. Real IPM is experimenting with (and planning to register) a campaign for fruit fly control. This will involve a delivery system for *Metarhizium* that uses a trap and lure to bring the fruit fly to the *Metarhizium*. Mango farmers in Ghana have reported up to 70 per cent loss through fruit fly infestation.

Capacity

On capacity the team has worked with the EPA (Environmental Protection Agency of Ghana) to refine their registration administration and a guide has been developed to assist others interested in registering a biological control agent in Ghana. The Ghanaian EPA was found to be a regulatory authority that was keen to get BCAs registered. While no new businesses have yet been created, the process is proceeding so well it may provide a main market for Kenya Biologics who have been struggling to get products registered for use in Kenya. There is also interest from Senegal and other West African countries. In addition, there is talk of BCA registration being expanded for use on other crops and the team have attempted to facilitate future registrations of BCAs.

As part of the capacity building of the regulators, EPA and PCPB staff attended a bio control conference in Switzerland to make them aware of European developments in both regulations and product development. Real IPM is also investigating bringing the original Stopstriga isolate (see above), that was intended to be registered in Kenya, back to Ghana as a commercial product. The isolate came originally from striga infested fields in North Ghana. The introduction of such an isolate will not meet regulatory opposition to undertaking field trials in Ghana. However Real IPM is nervous about over-extending their capabilities and resources in developing this product for the North Ghanaian farmer.

Linkages and partnerships

The more important of these are shown in the sections above.

Broad summary conclusions

What can we conclude from these projects? To begin with, it is very clear that they all demonstrate the complexity of putting research into use. And this complexity has little to do with the basic science involved. As one of the scientists stressed in an interview, the underlying science associated with seed priming is well understood. But this is only the start of the 'applications engineering' problem since what is then needed is the wide range of capacities and resources needed to turn this raw knowledge into development. In part this is a marketing issue as we have seen with the Real IPM Stopstriga story, but it also depends on type of product, geographical context, weather conditions, complementary information, regulation and many other factors. We shall deal with these and related issues in the concluding chapter, but at this stage it is worth highlighting the following:

a) In development terms progress has not been fast. Over a 2.5 year period one Best Bet has moved into the production of one of its projected outputs (but not the other as yet); one is about to launch production (and has aroused private investor interest); one is still laying the regulatory groundwork. This illustrates the complexities involved even where the underlying science is well understood. It also reflects the influence of seasonal factors such as growing periods and rainfall patterns.
b) In the projects examined, national regulations and their application have proved a significant constraint. This seems mainly due to a lack of suitable guidelines for bio control agent use. Countries still use guidelines designed for the use of chemical biocides and have difficulty making appropriate judgments. This issue, however, varies across countries with, in the BCA case, Ghanaian regulators proving much faster to adapt than the Kenyan equivalent.

Notes

1 Further details on this Best Bet may be found in Gildemacher and Mur (2012).
2 See http://www.tanzania.go.tz/agriculture.html.

3 See http://www.fao.org/ag/AGP/AGPC/doc/Counprof/kenya/Kenya.htm.
4 See http://www.researchintouse.com/nrk/RIUinfo/PF/CPP43.htm.
5 See relevant discussion of extension systems in Chapters 1 and 7.
6 See http://www.lancs.ac.uk/staff/wilsonk4/ARMYWEB/ARMYWEB.html for more complete review and further details.
7 Interview with R Day, CABI.
8 See Wilson *et al.* (2009) and Grzywacz *et al.* (2009).
9 Interview data.
10 Russell IPM's technical director did his PhD at NRI.
11 http://www.lancs.ac.uk/staff/wilsonk4/ARMYWEB/.
12 Success at registration level can open doors to further markets. For example Roger Day and Richard Musebe at CABI mention attractants to deal with the Asian fruit fly. The Kenya PCPB has produced a draft paper on pheromone registration which may soon be accessible.
13 The facility harvests the virus from infected caterpillars and then processes this for production and sale.
14 See http://www.aphis.usda.gov/publications/plant_health/content/printable_version/fs_phwitchweed.pdf.
15 One microbe (Metarhizium anisopliae, for the control of adult thrips) which it has commercialised had been heavily researched by ICIPE but never commercially applied. Now ICIPE gets 1 per cent 'royalty' from Real IPM's sale of the fungi.
16 Striga push pull/GAP projects (R8212 and R8449).
17 Using the website, when farmers fields are visited their crops/events can be monitored by a photograph and uploaded from a digital camera; through GPS technology their location and picture can be displayed on a map.
18 These technology packs (Mycoherbicide, instructions and information leaflets) are written in three languages – English, Swahili and Luo.
19 Information from Fresh Produce Exporters Association of Kenya (FPEAK). See www.fpeak.org.

5 NGO-led projects

Introduction

This chapter provides summary details of three further Best Bets, in this case projects in which NGOs played a major facilitating role but which have perhaps not proceeded as well or as far as initially expected. They deal with aquaculture, seed yam and rice, and like all the Best Bets, represent attempts to leverage private sector inputs into technology development. In each case, however, there have been operational problems. Partly this has had to do with delays in start up. In other cases the complexity of the technology has proved to be an obstacle. Again, the text summarises the rationale for the interventions, the RNRRS projects that provided their scientific base, their co-operating organisations, objectives, progress and issues arising over the period. Detailed analysis of outcomes and implications for policy will be dealt with in the final chapter.

Aquaculture development

Building services and supportive policy and increasing fish production, enhancing food security through the dissemination of improved tilapia strains and application of 'best practice' aquaculture techniques.

Area of intervention

Traditional small scale fishing has become increasingly marginalised both by overfishing and by the incursion of commercial companies. An obvious alternative is fish farming. Fish farming, or aquaculture, is a non-traditional technology that has until recently had limited adoption by the resource poor in sub-Saharan Africa. Governments, national agricultural research organisations (NAROs), NGOs and donors have sought to promote it as a sustainable technology that can be easily integrated into farming systems, and which has the potential to improve food security and incomes for rural populations. There are, however, a number of constraints to the expansion of aquaculture in most African countries. These include poor supply and distribution of quality fish seed and feed, poor access for producers to technical and management information, poorly developed rural markets, and a lack of credit provision from financial institutions. The most appropriate and successful extension method appears to be the on-farm participatory extension approach, which is currently used on a trial basis in several countries. The sustainability of the method however still needs to be demonstrated. There are also environmental concerns over hydrological disturbances that may occur when wetlands are altered to accommodate fishponds (i.e. management activities such as constructing embankments and/or drainage schemes, or importing water), the introduction of chemicals and pesticides that may enter the wetland as residue on fish feed, and biodiversity impacts on existing fish and wetland species.

Resources in Sub-Saharan countries have considerable untapped potential for aquaculture production, in particular clean water and unused land. Low-value food fish has emerged as an environmentally friendly production system that provides a cheap source of protein. Growing demand and expanding markets are expected to push fish prices up, hence the need to increase the supply of low-value food fish to keep the price within the reach of the rural and urban poor. Meanwhile, the supply of fish from capture fisheries (e.g. from Uganda's lakes and rivers) is threatened by increasing pollution, unsustainable fishing practices and the proliferation of water hyacinth. The situation of increasing demand and decreasing supply from catch fisheries provides an opportunity for smallholders willing to invest in aquaculture technologies. Starting from a small base, aquaculture has had rapid growth recently in Kenya, Uganda and Malawi. For example Uganda has had almost a 3,000 per cent increase since 1994. Further reasons for growth include growing urbanisation and income, expanding markets and services, improved skills, opportunities for private sector development, an increasing and food insecure rural population and new production technologies. Some experts have therefore concluded that aquaculture development has good potential in certain areas of Uganda.

Seed networks enable entrepreneurship among the poor, allowing them to be involved as small seed producers, nursery operators, traders or sellers of materials and supplies related to seed production and delivery. As networks expand, roles become more specialised. The diversity of livelihood niches tends to benefit even the more disadvantaged people and groups. Setting up 'one-stop-aqua-shops' (OAS) has proved to be a successful mechanism to boost productivity in various Asian countries. These are mobile shop/service providers selling products and providing services and up to date information to farmers. A World Bank report commenting on 'weak African extension services' in general has highlighted OAS as potentially responsive to the new challenges and opportunities, and the demands of farmers (World Bank, 2007: 51) while a related report summarises the benefits of an OAS approach (World Bank 2006b: 39, Box 5).

RNRRS projects

There were four of these, all having been carried out in East Asia:

- *R6759 – Integration of aquaculture into the farming systems of the eastern plateau of India*
- *R8100 – Investigating improved policy on aquaculture service provision to poor people*
- *R8334 – OASIS: the one-stop-aqua-shop*
- *R8363 – Enhancing development impact of process tools piloted in Eastern India.*

They covered the integration of aquaculture into poor farming systems through support services, policy reform and capacity building. The DFID RNRRS original research team had developed effective ways to encourage the private sector to drive the growth of the aquaculture sector and to create private-public partnerships in service provision called Aquashops. Large-scale uptake of aquaculture had brought widespread livelihood benefits for 20,000 poor farmers and entrepreneurs in eastern India. Highly effective Aquashops provided access to good quality seed, feed and technical advice (a process already replicated in India, Pakistan and Vietnam, and also adopted by UNDP and World Bank poverty alleviation efforts). In addition, through RNRRS (projects R8100 and R8334), the original researchers involved successfully supported farming and implemention of policies for commercialisation and privatisation, building a supportive policy environment with government and farmers through a replicable, participatory process of consultation and consensus-building. The main lessons learned were:

- Fish seed networks in different parts of Asia have grown to both encourage and then support the growth in aquaculture, including generating employment to the rural areas.
- Bringing seed to potential adopters of fish culture, often with information of how to grow them, has been a critical factor in the now widespread stocking of water bodies.
- The poor are largely involved in the seed marketing network as traders and service providers; thus any improvement of these networks is likely to impact on them. In addition, it has been beneficial to have close contact with government efforts to promote aquaculture and to develop the required infrastructure.
- A key aspect of any government support for private sector seed networks, particularly in regions with poor road infrastructure, is a commitment to decentralised approaches.

In short, the research conducted over ten years (www.nrsp.org/6_6_1.aspx) appeared to define processes and knowledge, suited to achieving significant development impact at scale in East Africa through working with the governments and investors in Kenya and Uganda.

The project

The following bodies were involved in this project:

a) The consortium was led from Nairobi by *Farm Africa (FA)* who have championed and popularised the Aquashop approach, facilitating linkages, building on its animal healthcare model and drawing on experience administering the Maendeleo Agricultural Technology Fund, to manage start-up support to selected Aqua Shop franchisees.
b) A specialist consultancy firm, *Imani Development (ID)* provided Aquashop franchise owners and smallholders with business management support and conducted market research.
c) The *Kenya Agricultural Research Institute (KARI)* promoted Aquashops (including in the semi-arid areas) and helped to provide a supportive national policy environment.
d) *The Original Researchers from Stirling University* are specialists in aquaculture in development contexts. They steered the initiative in its early stages, applying tacit knowledge and learning, linking with producer organisations and facilitating a policy change process.

The business model was as follows:

- To identify a franchisor with the capacity to establish and support a network of franchisees in Kenya.
- The franchisor would establish an Aquashop in Kenya to serve as a model for recruiting franchisees.
- The franchisor would recruit two types of franchisee – one for areas with high aquaculture potential and the other those where potential is lower but the benefit to local livelihoods and food security needs high.
- The high potential area franchisees would operate in more accessible areas, with more established markets, thus generating higher potential incomes. They would have a social impact through providing quality services at affordable prices.
- The low potential area franchisees would operate in more remote areas, or new markets, providing much of the direct social and economic impact of the business.
- There would be up to six franchisees in Kenya servicing six out-growers each, with approximately 15 members each, supporting in total up to 1,000 out-growers.

The project would disseminate best aquaculture development practices in relevant languages, consolidate existing research, improve access to information and links amongst out-growers, private hatcheries, quality seed and fingerlings suppliers to prospective and current fish farmers in Kenya. In so doing it would understand the likely policy, structural and regulatory changes to be created or revised to improve opportunities for commercial and small-scale aquaculture development. These components and/objectives were expected to complement the Kenya Ministry of Fisheries initiative under the Fish Farming Enterprise & Productivity Programme (FFE & PP). It was anticipated that the emergent Aquashop franchisees, after receiving adequate support and capacity building, would continue operating as successful, commercially viable, stand-alone enterprises.

Progress

Techno-economic

This Best Bet was late in being started and the launch of the six Aquashops did not take place until February 2011, since this required the previous completion of a market study to finalise and promote the Aquashops concept and set up links with service and goods suppliers. Effectively therefore, the planned Phase 1 of the project did not begin until July 2011 and there is little direct evidence yet of economic progress. Systems to develop the sector were set in motion by providing support to a private sector firm, Jewlett Enterprise, to scale up quality seed multiplication. Jewlett were initially selected because they were seen to be young, aspiring entrepreneurs (recent graduates of Moi University) with some experience working for Dominion Ltd., the main seed supplier, and some consultancy work on selective fish breeding. The enterprise also appeared to have potential to become a franchisor in the future. Potential franchisees were then asked to submit bids and after vetting and interview, six best candidates were picked. For a franchisor to operate profitably, high volumes of inputs and supplies must be moved. Availability of quality seeds at the right time is key to the sub sectors' growth. Support to Jewlett Enterprises to produce quality seeds was to address persistent unavailability.

During the last week of March 2011 the team conducted a field survey in Samia. The Samia survey showed that out of the 31 people questioned only 7 (approximately 23 per cent) were women. Apart from fish farming being an alternative source of livelihood, the women stated that fish farming is less time consuming giving more time to attend to other issues. However, traditional beliefs over land ownership (male dominated) continue to affect the participation of women in fish farming. Further work in processing and marketing is expected to include more women and youth at the upper end of the value chain where the basic factors of production that contribute to their exclusivity are not necessary. Approximately 28 per cent of the project beneficiaries were women, the rest men. The survey also showed that potential economic benefits are considerable but patchy. Some farmers earned as much as Ksh 150,000 (€1500) in one production cycle, while others earned as little as Ksh 2,000 (€20) in a similar facility of the same measurement; this indicates how factors such as management, the level of commitment and the ready availability of inputs are often key to success. Overall, there are a range of issues confronting the development of this sector in general and Aquashops in particular. These are:

- A great shortage of suitable seed fish. Only one company has the capacity at present, though two or three are in the pipeline. It is reluctant to scale up because of the 'one off' nature of the business.
- There are unexplored issues of fingerling quality since the biology is complex and further research needs to be done.
- There has been a tendency to over-specify feed quality since artisanal fish farming can be done with fish feeding on natural organisms in ponds (at least to some extent).
- In any case there is a severe shortage of pelletised feed since feed suppliers are reluctant to commit themselves in an uncertain market climate.
- There are technical issues of how best feed should be introduced into ponds and on 'one sex' fish farming that need to be clarified.
- Prices of fish feed are still very high for the small scale fish farmer. This tends to put off many farmers, although to address this, Aquashops are breaking the feed into small packages.
- Drought experienced in recent years has forced some farmers to abandon their ponds. However, there is evidence that the majority of farmers are now giving orders to Aquashops for seed.
- There are still issues about the viability of the franchise model due to diseconomies of scale and scope.

Capacity

During the project's early phase various studies were undertaken, e.g. Imani's market study and an information needs assessment which led to the development, sourcing, translating, adapting and testing of information packages. This meant that best practice was shared with each batch of franchisees through information packages, and a system of ongoing M&E, review, and revision of information packages was put in place. Training modules on business management and Aquaculture services and products were developed and collated into an Aquashops *Operators Training Curriculum*. The development of this curriculum was not a planned output but became necessary as an essential standard and guideline material for future use and sustainability. As well as developing the training modules themselves, the training of the franchisees in business management and aquaculture services was achieved. In terms of beneficiary groups 552 farmers attended project training programmes; 6 franchisees and 3,153 farmers benefited indirectly from Aquashop services. It was also expected that 12,000 fish farmer households, 30 employees on the farm, and 6 Aquashops would benefit from seeds from the franchisor; this was a local company called Jewlett Enterprises.

An *Aquashops Operators Association* was formed and farmer groups created to provide support and marketing services to members. Operational guidelines and standards for the establishment, set up, storage, display facilities and minimum technical qualifications required to run the Aquashops were developed. Farmer group support through training in best fish farm management practices, and organisational development was provided to help improve on farm management by farmers, strengthen clusters and to stimulate demand for Aquashop services. The team developed an Aquashop logo, used for branding the Aquashops for ease of Aquashop identification and also as a marketing strategy for the Aquashop services and products.

Linkages/partnerships

With the assistance of Kenyan universities, Aquashop franchisees were trained both in issues affecting fish farmers and also in business. A syllabus was developed that could be used by others and hopefully this also helped to strengthen the university sector. A comic story was run by WTS Ltd. (see Chapter 6). In response to the Shujaaz story, Aquashops received numerous calls from readers. Most of them asked how to start fish farming and how to acquire quality inputs. Over a month after publication these calls were still coming through and the team used the opportunity to assist them. The Aquashop team feel that through the Shujaaz story they have managed to reach a substantial number of interested members of the public and they fully support this approach. They also visited Aquashops in Orissa and were impressed by the number of beneficiaries and the technology being used, e.g. portable hatcheries. The University of Stirling produced six Aquaculture Best Practice Information factsheets.

The project has also worked to catalyse policy change at national government level; this included for example setting up an initial stakeholders' consultative meeting to bring understanding on the likely policy, structural and regulatory changes necessary for aquaculture development in Kenya; engaging in a consensus building process with policy players resulting in policy change priorities; submission and follow-up of findings, and policy change priorities with government decision making bodies. A draft regulation on licensing of Aquashops was developed and subjected to stakeholders' validation before it was gazetted. A geo-referencing mapping exercise was undertaken to identify the most viable Aquashop locations vis-à-vis surrounding fish farms/ponds undertaken.

As of 2009, the GoK had included fish farming as a key growth sector. Since then it has funded some 48,000 ponds in 160 (out of 210) constituencies. Finance has come from the Treasury to the Ministry, with pond construction (10 labourers per pond) paid for by the Kenya Community Development Fund (CDF); 1,000 fingerlings per pond were supplied free. The Permanent Secretary of Ministry of Fisheries Development spoke to the Best Bet team leader before the Aquashops launch and he expressed the view that the Ministry would like Aquashops in all the 160 constituencies where FFE & PP is being implemented. The Director of Fisheries has also maintained that the Aquashops initiative is an excellent exit strategy for the ministry after the FFE & PP ends.

Seed yam

Area of intervention

Worldwide yam production in 2007 amounted to 52 million tons, of which Africa produced 96 per cent. Most of the world's production comes from West Africa representing 94 per cent, with Nigeria alone producing 71 per cent, equalling more than 37 million tons. African countries imported more than 2,000 tons in 2002, and exported 15,500, of which Nigeria exported 12 per cent. More than 95 per cent (2.8 million ha) of the current global area under yam cultivation is in sub-Saharan Africa, where mean gross yields are 10 t/ha (http://www.iita.org/yam), Yam (*Dioscorea spp.*), especially white yam (*Dioscorea rotundata)* is a highly valued crop (economically, nutritionally and culturally) in Nigeria and throughout West Africa. There is huge demand for yam consumption from both domestic and export markets; however the full potential for income generation both through domestic markets and the export trade has not been realised due to problems and inefficiencies in production, handling and trading systems (IITA).

Yam production is in decline and much of this is due to seed borne diseases exacerbated by the fact that planting material is scarce. An expanding human population means increased pressure on the land to produce more food, resulting in shorter fallow periods which leads to a reduction in soil fertility and the build up of pests and pathogens in the soil. Yam is thus faced with increasing levels of field and storage pests and diseases associated with intensification of cultivation, and certain yam varieties are disappearing. Disease free seed yam can be produced but it takes time, resources, expertise and traditional credit, which is very expensive. The market for clean seed yam is weak with retailers disadvantaged through market cartels which limit private sector operations. One of the big issues with yam is that it rarely produces seed. Instead yams are vegetatively propagated through planting pieces of yam tuber. But, diseases can be carried over from one crop to the next in the tubers, and productivity remains low. Another factor is that the planting material costs for yam are extremely high, as much as 70 per cent of production costs in some (Ampofo *et al.*, 2010) and even then there is no guarantee in place that the yam will be disease free (DFID). What is needed is an affordable source of 'clean' or disease-free planting material of the varieties they prefer.

One option is for farmers to grow their own disease-free material. The 'minisett technique' was developed over 30 years ago as a means of rapidly multiplying yam. It is a two-stage process that produces many small 'seed' yams in the first year, which can then be grown in the second year to produce larger 'ware' yams suitable for eating and trade. Researchers have been working with farmers in Nigeria to adapt the technique to farmers' needs (Kenyon *et al.*, 2005).[1] This is a method that has managed to keep costs to a minimum and any environmental hazards from the pesticides is kept low, while still breaking the disease cycle. The practical adoption of research findings arising from these projects (largely funded by DFID and IITA) could go a long way to ensuring the production of the crop mainly through the provision and supply of planting material that is stable, predictable and affordable.

RNRRS projects

This Best Bet was built upon an extensive history of RNRRS research which has led to a greater understanding of yams and improvements for yam producers. This research has sought to understand the epidemiology and biology of yams themselves and the epidemiology, biology and control of yam diseases such as anthracnose, and with this knowledge develop control strategies. Other diseases, for example yam rot and fungal rot pathogens, were investigated and had the causes of the disease identified; pests that attack during storage were also identified. Capacity was increased through establishing laboratories and training a local technician. Other research provided new knowledge to initiate improved control practices (clean seed tubers, rotation where possible and intercropping with maize). Local capability in plant pathology had been enhanced through inputs to laboratory and through training. Recommendations for improved and sustainable pest management practices were formulated and tested on-station, and if successful were promoted to smallholder farmers.

Further work was done on promotion of yam technology with research that focused on a) evaluating crop protection practices (based on current practices, local technical knowledge and the outputs from previous projects) for clean seed yam production for applicability and economic efficacy in Kogi and Ekiti states of Nigeria, and b) developing dissemination outputs related to clean seed yam production which would lead to improved seed yam health and availability through promotion within Nigeria initially; it would also be suitable for wider application across the yam-growing belt of West Africa. In all, a thorough exploration of the systems and economics of seed yam production and supply had been carried out, and the systems developed to produce good quality seed yams had been evaluated, demonstrated and promoted. In summary the following projects were funded by DFID:

- *R5259 – An examination of Dioscorea spp (yam) for nematode resistance and its incorporation into improved cultivars*
- *R5345 – Epidemiology and control of yam anthracnose*
- *R5346 – Biology of yam anthracnose (Colletotrichum)*
- *R5688 – Natural resistance to infection by Colletotrichum*
- *R5983 – Factors influencing the occurrence of yam tuber rots in West Africa*
- *R5738 – Epidemiology and control of anthracnose disease of yam in Nigeria*
- *R6691 – Control of yam diseases in forest margins farming systems in Ghana*
- *R6694 – Identification of resistance to major nematode pests of yams in West Africa*
- *R8278 – Evaluation and promotion of crop protection practices for 'clean' seed yam production systems in central Nigeria*
- *R8416 – Upscaling sustainable clean seed yam production systems for small-scale growers in Nigeria.*

The project

The consortium had four members:

- Diocesan Development Services (DDS). DDS was the lead partner in the project. Responsibilities included the production of clean seed yam on its Iyegu farm, identification and support of farmers to apply the technique on their own farms, training of farmers, and contact with media. Named individuals were Sister Felicitas Ogbodo and DDS staff.
- Congregational Development Office, Missionary Sisters of the Holy Rosary. Acting as interlocutor of finances and reports. The named individual here was Sister Nora McNamara.
- IITA. Technical support, especially with regard to processes for assessing the quality of seed yam. Named individual was Dr Robert Asiedu.
- University of Surrey (Centre for Environmental Strategy). Technical, training and managerial support for the activities in Nigeria. The named individual here was Professor Stephen Morse.

DDS and others had already been active in seed yam projects. In 2010 it became part of a national programme, Global Food Security Response

Initiative, a commissioned project of CORAF/WECARD, promoting clean seed yam technology. IITA was implementing this project in collaboration with a team of national partners in 13 research institutions in six countries, including Nigeria's National Root Crops Research Institute (NRCRI) in Abia State in south-eastern Nigeria. As part of this project over 160,000 seed yam were sown on the DDS seed farm and by farmers throughout Igalaland to promote the clean seed yam technology. This national programme acted as the precursor of the DFID-RIU initiative and offered a unique opportunity to bring together the farmers from many yam growing areas so that they could avail themselves of suitable technologies available. The means of production would be within the community rather than having to source planting material (often not adequate or suitable) many miles from operation (in this case Igalaland). Also a strong farmer's organisation in Igalaland, facilitated by DDS, would provide a useful user springboard.

DFID-funded research had analysed yam systems in West Africa and had identified that pest free 'clean' seed yam is the key to unlocking many of the technical constraints. In Nigeria there were a small number of efficient seed yam production systems but these were mainly operating at a community level. The financial, social and cultural benefits of more efficient yam production could only be achieved when these systems operated at a larger scale. This RIU initiative was therefore about carrying forward the knowledge previously generated to improve the efficiency of yam systems in Kogi and Benue States so as to fulfil the development potential of clean seed research and to demonstrate the wider development potential, particularly for scaling up of this approach through other development initiatives.

In more detail the project would bring together the private, public and NGO organisations that would make yam systems more efficient in Kogi and Benue States. As well as exploring the key constraints and identifying solutions, these organisations would also be invited to be kept informed of practical actions that are taking place at a farm level amongst 250 household and other groups. They would be facilitated to produce high quality disease free (clean) yams. It was also planned to demonstrate how best to select planting material from the early and later harvested yams so that smaller growers would be in control of producing their own supplies of clean seed yam. Larger growers would then be identified who could act as centres of clean seed yam production. This group would hopefully include business people but in any case these growers would eventually become commercial suppliers of clean seed yams for other farmers given the assistance of the proposed DFID project.

Methods of clean seed yam production would be applied at the Seed Multiplication farm located near Iyegu village, where skilled trainers and facilities were available. This would provide both a demonstration and at the same time a start-up source of material for the farmers. Training would also be extended to the village level where there were skilled yam farmers. Some of these knew about Minisett technology. The farmers would help facilitate this training with support from DDS, and would include the more marginalised in society. Training would also embrace schools (especially those which are faith-based) and colleges. Storage and soil fertility have long been identified as being critical to the health of seed yams. Education and training in these aspects would also take place at the Seed Multiplication farm at Iyegu, where there are improved storage facilities, and in the villages and schools. Wider promotion in relation to clean seed yam production and storage was planned through publicity in the media mainly through radio and television as well as through posters and pamphlets. Links would be maintained with funding from the BMGF so that this work would provide the necessary lessons for their scaling up of this approach.

Progress

Techno-economic

The RIU yam team have completed training in the production of healthy seed yam for all of the 120 communities covered by this project. When the rains began, ten communities received material assistance through treated seed yam for planting. Film was produced for the RIU which would be used to highlight the approach and stimulate further networking. There are now 250 households using the clean seed yam technology and DDS are looking to increase this by 300 HH in 2012, and 3000 over the next 4 to 5 years. In addition there has been the establishment of 20 seed yam producers as entrepreneurs trained in good agricultural and business practice along with the provision of credit through DDS to seed yam entrepreneurs. In fact DDS have decided that a seed yam business would help them to raise funds and so have allocated resources to join this project as a seed yam business. Finally, 8,000 tonnes of ware yam have been produced which has benefitted 131,000 people.

Male farmers in Kogi State have informed the team that the cultural constraint which prohibits women from directly engaging with crop production is now being challenged. Previously women were restricted to overseeing the crop in storage. Diseases (originally soil borne) can cause major post-harvest losses, leaving the family with insufficient funds for school fees. The link between pre- and post-harvest losses shown to farmers in this work has resulted in farmers realising the benefits of women being involved in the whole cropping process, including the more lucrative process of land preparation. This change in attitude undoubtedly will have a carry over to trading, for if women have the opportunity to grow and market clean seed yam this would impact on income and set the scene for poverty reduction or elimination. This is a new situation which could be nurtured and spread.

Yam farmers trained in seed yam production under the RNRRS are continuing to produce seed yam. The demand for disease free seed yam is greater than it was at the end of the RNRRS. However, the farmers have failed to increase production levels for a number of reasons, including a lack of business skills and business services. Finance is a major constraint; the NGO and DDS, has a credit scheme which is available to yam farmers. Interestingly, although this credit scheme is well established, the yam Best Bet is bringing about a change of mindset amongst credit receivers who now see the need for it as an investment rather than as something to tide them over during the times of year when they have the least money. It has also been discovered that as a result of training farmers under the RNRRS there has been a level of copying the approach by neighbouring farmers; these farmers are also not generally producing large quantities for market but they show potential for scaling up the approach.

Capacity

As with Aquashops, this Best Bet was delayed in start up for weather/planting reasons, in this case to October 2011. Therefore, main activity has focused on providing business training and additional technical support to seed yam producers so that during the 2011 season they could become seed yam entrepreneurs. The Best Bet is operating in six states and has expanded to include schools and other development projects as well as seed yam producers. The main reason for the lack of expansion of scaling up seed yam production was the lack of credit. Kogi State has weak service provision and so the financial services offered by DDS have become a critical source of capital to yam farmers. The loan provision itself has become more business orientated (requiring a business plan) and is therefore complementary to the RIU Best Bet. The farmers have also received business training and support and have developed a business plan model which would be implemented during the forthcoming rains. There is a lot of enthusiasm amongst project staff and participating farmers. Farmers in Edeke (Kogi State) have already planted the seed yams and it looks as though the farmers appreciated the training provided and have all adopted the training recommendations.

Partnerships/linkages

In Abuja (FCT) the project links with the MDGs, a Federal Government Project whose aim is to train farmers, especially women and youth, in crop production and processing. In Rivers State the Best Bet links with Food for All International (FFAI), an NGO. Many farmers were trained but only ten could be provided with minisetts for planting (2,000 – half from the RIU and half from the NGO). The team are also working with Rivers State Prison Service providing training for 12 officials from their farm and

28 inmates: 2,000 treated minisetts have already been planted. The team have also been working in partnership with the Green River Project and supported a field day which was attended by 200 farmers, five of which were identified to receive material support from the Best Bet.

Transfer and dissemination of emerging agricultural technologies of new rice for Africa (NERICA): improving access to quality seed through public-private partnership in Uganda

Area of intervention

Rice is relatively new to Uganda but production is growing fast and there is a shift from other cereals such as maize, millet and sorghum to rice. Though the increase in rice production in Uganda has been very substantial, the country is still not self-sufficient. At project inception total rice consumption was estimated at 210,000 metric tons with local production of 144,000 metric tonnes and the rest being imported, at a value of US$30–50 million annually (RIU data). Because of the huge domestic demand and the fact that rice gives a higher rate of return on investment (output to input ratio of 1.83 compared with hybrid maize and sorghum) this crop's contribution to agriculture is undoubtedly enormous. The total population of Uganda is 28 million with an annual growth rate of 3.2 per cent indicating that rice consumption is likely to increase.[2] Rice is an emerging priority crop in the Government of Uganda strategies because of its potential to greatly reduce household hunger and poverty. Among staple food crops, rice represents the country's best opportunity for reduction of imports. It is also a major source of employment for the rural poor. Its marketing and processing represent a boost to development of rural centres through installation of facilities such as electricity and milling machines.

Earlier emphasis was given to irrigated rice but, from 1990, government priority shifted to upland rice because of environmental concerns on the continued use of wetlands for rice production. This alternative was considered achievable, considering that about 70 per cent of the country is suitable for double cropping of rain fed rice. Uganda initiated a programme to look for improved varieties and from this a total of 110 varieties and breeding lines were obtained from the Africa Rice Centre (WARDA). These were tested on-farm in the 1999 rainy season to identify the best ten varieties which were then advanced for on-farm and on-station evaluation in five key locations. From these studies the NERICA varieties of rice performed very well (Hirvonen, 2011).

NERICA rice varieties with high yields and relatively low water requirements are seen to support government initiatives. This Best Bet responds to concerns that initiatives focusing on grain production as a means to support improved smallholder livelihoods are hampered by low availability and poor quality of NERICA seed. The private seed sector in Uganda is not well developed; companies generally operate on a relatively small scale and rely on outgrowers to augment their seed production, however, they do not have the capital to invest in training farmers to increase production of high quality seed to meet the demand from farmers. As a result, companies are forced to accept seed purities lower than the 98 per cent specified in standards as well as a high risk of failure of farmers to deliver. They are also having difficulties in increasing volumes of seed produced. Basic seed is provided by the NACRRI seed unit which has mechanisms in place allowing it to operate commercially. However, currently seed is sold at a subsidised rate because seed companies are reluctant to pay the full price given the uncertainties they face at the bulking stage. Three varieties of NERICA rice have been registered by the Ugandan National Seed Certification Service. These are NERICA 1, 4 and 10. Two local seed companies, NASECO and FICA Seeds, have exclusivity contracts for 1 and 10 respectively. The Best Bet is working with NERICA 4 which is open access.

RNRRS projects

There were four RNRRS projects used in this Best Bet as follows. It should be noted that three of these projects dealt with generic seed issues producing knowledge applicable rice production. In addition NERICA research was also informed by WARDA funded through a basket of international donors.

- *R7251 – Promotion of pro-poor strategies to reduce the impact of key pests and diseases in vegetable crops*
- *R8312 – Promotion of quality vegetable seed in Kenya (2003–05)*
- *R8439 – Promotion of quality kale seed in Kenya*
- *R8480 – (2005–06) Good Seed Initiative (GSI).*

Project

The following groups formed the consortium:

- Dr Godfrey Asea; National Crops Resources Research Institute (NACRRI). NACRRI has coordinated and led field activities in Uganda and has provided basic seed for bulking commercial seed by the seed companies.
- N. Rodeyns, NASECO 1996 Ltd. Seed companies are key partners in the consortium as they will continue to work on NERICA rice seed production beyond the end of the project. NASECO has established direct relationships with out-growers contracted to deliver seed. Company staff supported supervision of farmer groups and learned

how to train farmers, so that they can expand activities in the future. Distribution of seed has been largely through existing seed company networks and production complemented by the companies' own production on owned or rented land.
- T. Baligeya, Centre for Agricultural Industry (CAII), a smaller private sector firm.
- N Phiri, CABI-Africa. CABI has provided support in terms of planning and designing training modules, working closely with NACRRI. They have also led in development of information materials to be used in the training programmes, and worked closely with seed companies and NACRRI to ensure that materials are designed to provide the information needed by different stakeholders.

The project aimed to support scaling up, production and delivery of high quality NERICA rice seed by local seed companies and to implement a marketing strategy to stabilise demand. It supported the costs of training out-growers and establishing new farmer groups to work with seed companies, while at the same time training company staff as trainers so they can continue activities in the future. Training materials would be produced and made available for use by other companies and entrepreneurs wanting to engage more effectively with out-growers. Promotional activities, including use of radio as well as other information materials such as posters and product flyers, would stabilise demand through creating awareness both of the value of the product and of the importance of using high quality seed. In summary, the project had three broad objectives:

1) To increase NACRRI capacity for basic seed production to 5 MT/year;
2) To increase capacity of targeted seed companies to produce 1,400 MT by establishing functional linkages with out-growers; and
3) To stabilise product demand.

Progress

Techno-economic

The essential steps in this project were as follows. Firstly, the Africa Rice Centre (former WARDA) supplied foundation seed for the different varieties of NERICA rice, and NACRRI evaluated these for suitability in Uganda. Suitable ones were registered with the Ministry of Agriculture, Animal Industry and Fisheries so that they could be legally grown in Uganda. The basic seeds chosen were of NERICA 1, 4 and 10. These were cleaned up (by growing them and removing off-types) and supplied to the two seed companies as Foundation seed 1 (basic seed) to produce Foundation seed 2 on company land. The Foundation seed (2) was then given to contract farmers for production of certified seed. The companies enlisted the out-growers who formed groups. By early 2012, the total number of contract farmers registered was 275. The companies contracted for set re-purchase buyback orders with multiplication rates that proved to be upwards of 60:1.

NACRRI established 5 ha of foundation seed for all the three varieties and 2 ha of breeder seed from the WARDA sourced nucleus seed. The NACRRI target was to produce 5 MT of seed per year. However in practice too much seed was produced (7.5 tonnes) due to additional support from the government of Uganda (probably due to increased demand for rice). The farmers then sold the excess NERICA rice seed as grain for milling rather than waiting for the next growing period. The seed companies were also direct beneficiaries of this.

During 2010, NASECO established 25.5 acres of NERICA 10, 391 acres of NERICA 4 and 62 acres of Suparica 1. The seed farm established 25 acres of NERICA 10. Overall the companies produced 857.1 MT of rice seed in 2010, less than predicted due mainly to drought. The direct beneficiaries were 230 contracted farmers, and indirectly over 100 agro-input dealers, transporters and packing material dealers. Other indirect beneficiaries were the farmers who would be buying and growing the seed for food. So far NASECO has sold 218 tonnes, while CAII has sold 10 tonnes. The low tonnage sold by CAII is attributed to drought which hit the rice seed crop before flowering. The market share of direct sale to farmers through agro-dealers has increased but no direct information could be captured from agro-dealers; however, on contacting one of these (NASECO), its CEO indicated that their seed sales had increased by about 30 per cent from 300 MT to about 400 MT in the first 2011 seed crop harvest. For CAII, their Managing Director reported a rise from 15 MT to 20 MT.

There was a rise in production area from 45,000 ha in 2009 to 53,000 ha in 2010. Rice grain production increased from 150,000 MT in 2009 to 180,000 MT in 2010. The 2011 production level was 195,000 MT, with the increase in production being mainly attributed to increased availability of quality rice seed and farmer awareness about proper management practices brought about by the project activities. This represents a considerable improvement over a two year period indicating that Uganda is nearly self-sufficient in rice. The Best Bet would have liked a continuation of funding beyond June 2011 for two purposes, (i) getting 'branded' rice into supermarkets and (ii) conducting a proper survey on the socio-economic impact of the Best Bet.

Capacity

CABI and NACRRI carried out training of trainers for seed company staff, who in turn trained farmer group leaders. Further training involved a practical aspect in which CABI, NACRRI and company staff trained farmer group leaders and farmers at central group plots through a participatory Farmer Field School approach (FFS) where farmers learn by doing. They subsequently transferred to their farms what they were learning from the FFS plots. This FFS approach is making farmers more willing to try out new technologies (seed and rice production technologies) by applying what they learn from the study plot. Eleven farmer groups participated in FFS for CAII. Each farmer group comprised some 25 group members on average thus giving a minimum of 275 farmers trained for NERICA rice seed production. As a result of FFS training, 20 farmers have planted rice for seed on their individual farms, each averaging 0.5 acres.

In slightly more detail, farmers were trained on timely harvesting, threshing, drying, seed storage, site selection, land preparation, marking and planting. In addition the project developed information materials for farmers, extension and agro-dealers to provide details of management practices, costs and benefits as well as to raise awareness of the value of good quality seed and good seed management. This included translation of key materials, including videos on 'Rice Seed Health', into local languages and preparation of content for magazines and radio programmes involving key stakeholders (farmers, extension workers, researchers, private seed companies, millers and representatives of regulatory agencies). Quality rice production manuals were developed, published and distributed to stakeholders in the rice seed value chain.

On social inclusion the NERICA Best Bet project embraced all rice seed farmers irrespective of their gender. Its records reveal that the proportion of women among the participating farmers is equal to that of men. However, by the nature of the crop, most of the labour is provided by women for planting, weeding, rouging, harvesting, threshing and winnowing, while the youth are involved in transporting and guarding rice crop against birds during the grain filling stage. It is therefore important that during training at these crop stages women are involved. Training also appeared to indicate that groups with a majority of women have been more committed.

Linkages/partnerships

There have been a variety of these developing over the course of the project. The RIU supported the March 2011 visit of the Ugandan seed rice team (team leader, researcher and seed company representative) to the RIU Zambia programme. This led to brand name development while the Zambians were able to share their experiences on rice marketing and benefitted from the Ugandan expertise on seed production. The Japanese aid agency JICA provided an infrastructure grant to NACRRI which will enable on-station training to key farmers on practical aspects of rice seed production. JICA is completing construction of a big regional (Eastern and Southern) rice training facility. In addition, the zonal agricultural research and development institutes have been trained and empowered as trainers in each zone. The project works closely with staff in these institutes where the project area fails to train and reach more farmers.

AGRA have co-funded the airing of radio programmes in the areas where CAII operate. The team were also in contact with the Department of Crop Production and the Rice Steering Committee through project updates in meetings and feedback from rice seed videos. The Uganda National Agricultural Advisory Services (NAADS) staff use information materials developed from the project. There has been acceptance by NAADS of information materials produced by the Best Bet, and they are now being used in their programmes away from project areas. A further unanticipated outcome was a request from ASARECA to out-scale the project to North Uganda and South Sudan (the then Southern Sudan). However, although activities have started well in Northern Uganda, activities in South Sudan will start only when the country finalises its membership to ASARECA.[3] At government level policy makers have been engaged in activities of seed production and developing some of the information materials. The National Rice Development Strategy provides a framework for coordinated activities for enhanced rice research and production in the country. Quality rice seed is emphasised as one of the drivers for increased rice production to achieve self-sufficiency. Policy makers were therefore engaged through the National Rice Development Strategy and the National Agriculture Advisory Service, both part of the Ministry of Agriculture.

Broad summary conclusions

As mentioned above, each of these projects relied strongly on initial NGO input, although like all the Best Bets they had the primary aim of leveraging private sector activity as well. Here the outcomes have so far been moderate. In the Aquashop case there are clearly some doubts about the franchise model and whether this is the best way to encourage and support technology development at farmer level. There are also issues associated with the complexity of the technology and the availability of necessary inputs. In the case of seed yam it has so far proved difficult to establish viable seed yam entrepreneurs while in the rice case the jury is still out on the sustainability of the relevant supply chain, though the out-grower system appears to have worked very well and Uganda as a whole is approaching self-sufficiency.

Nevertheless, there are signs that given more time things could improve. Indeed, what is common to each project is economic potential combined with the need for more time to support technology development. In the Aquashop and seed yam cases, projects have been late in starting due to exceptional weather conditions but in all cases it is also clear that projects of this kind need resources to build financial and technical capacity, and stakeholder awareness including buy-in on the part of government. Probably another year of funding would have allowed sustainable development of the NERICA rice value chain including the establishment of a brand name at agri-business level while the decision of the BMGF to take over support for the seed yam project in a much more intensive way is a testament to its long term viability. Technological complexity is indicated by the continuing involvement of scientists (including, importantly, those originally part of the RNRRS programme) in two of the projects.[4] It also indicates that technology development is a continuous process requiring technical applications input. Finally it is likely that there is the need for a continued supportive role on the part of both government and overseas aid.

Notes

1 See also Eyitayo *et al.* (2011) for more recent material.
2 These data refer to the situation as of late 2008. See www.unffe.org/index.php?option=com_content&view.
3 No information at time of going to press.
4 The University of Stirling input into the Aquashop project was discontinued in July 2011 following the departure of the scientists involved.

6 Other projects

Introduction

This chapter summarises the remaining three Best Bet projects. Two of them were initially a joint enterprise between two organisations, FIPS Africa Ltd and Well Told Story Ltd. These organisations applied together for RIU funding and for this reason they are highlighted together in this chapter. However, for a number of operational reasons it was decided to treat them as separate projects since, as will become clear, the nature of their respective interventions and impacts have been qualitatively quite different.[1] The third project on stamping out sleeping sickness (SOS) is the continuation of an older programme running since 2006. As in Chapters 4 and 5, the projects are summarised and then brief concluding commentaries are provided.

FIPS: Empowering millions of small-holder farmers throughout East Africa to put research into use – a private sector-led extension service to address climatic threats to food security

Area of intervention

In Africa there are urgent calls for improved food production as poverty is growing and productivity is declining on the continent. Kenya, Tanzania and Uganda are all agriculture-based economies, with smallholder farming accounting for about 75 per cent of agricultural production and over 75 per cent of employment. However, contributions of smallholder farming, and agriculture in general, to the region's recent rapid growth during 2005–2008 have remained limited. Instead, growth has been driven by services, in particular trade. Many smallholder farmers experience low agricultural productivity because of low and unreliable rainfall (exacerbated by climate change), poor soil fertility, late-maturing crop varieties which are susceptible to disease and drought, and poor soil and crop management. It is thought that measures needed to improve productivity of smallholder farmers include (amongst others) training to enhance skills and to encourage technology adoption and innovation.

Technology is available to increase productivity, but the appropriate inputs are often not locally accessible; nor are they available in affordable quantities and there is a chronic lack of information or advice at the village level. There are approximately 15 million small-holder farmer families in the East African region who are affected by these constraints. They typically farm less than 1 ha of land without improved drought- and disease-tolerant crop varieties and use little or no fertiliser. Consequently, crop yields are rarely sufficient to meet direct family needs. FIPS directly addresses these problems by making appropriate farm inputs available, in small affordable quantities, and ensuring information on their usage is locally available in ten small-holder farmer locations in Kenya, Tanzania and Uganda with a work programme designed to benefit large numbers of poor people as quickly as possible.

FIPS is a not-for-profit company based in Nairobi, Kenya, largely donor-funded but also generating a small part of its own income; its mission is to improve the agricultural productivity and livelihoods of small-scale farmers in SSA.[2] Here marginal farmers produce very little on small farms (<1 ha), working on impoverished soils, using local varieties susceptible to disease (including Maize Streak Virus and Cassava Mosaic Virus), and drought. Improved crop varieties are available at local research stations but farmers are generally unable to access them. Farmers often grow crops which are poorly suited to their environment, e.g. maize, which is often grown where there is not sufficient water or soil fertility. Poor farmers are unable to afford most farm inputs (in the standard retailed quantities), which in turn constrains the extent to which they are marketed in marginal areas. To address this issue FIPS-Africa has developed and implemented an innovative methodology to improve small farmers' access to the appropriate farm inputs, and information on best use. Its aim is to reach as many farmers as possible as quickly and cost-effectively as possible. To this end, it works closely with scientists in agricultural research centres and private sector seed and fertiliser companies. Close links with the private sector ensure continued supply of farm inputs after funding has ceased. Essentially, FIPS operates the following activities:

a) *Input supply*: This involves working with private sector input suppliers and retailers, the research community and private extension providers to promote the sale of small packs of affordable non-subsidised farm inputs, to develop the supply chain, and to make sure that appropriate inputs are accessible to farmers. FIPS-Africa's small packs are very cheap (e.g. tomato seed for Ksh 1) and are affordable to even very poor farmers enabling them to experiment with very limited risk.
b) *Technical advice*: FIPS-Africa develops networks of village-based agricultural advisers (VBAs) in target districts. These advisers are recruited from a variety of backgrounds but all have demonstrated the capacity to build trust relations with local farming communities.[3] Each

adviser works in five villages with a target of 1,000 farmers/year, although some go well beyond this. FIPS-Africa offers a range of services including: conducting demonstrations, selling small packs of seed and fertiliser, holding field days, establishing bulking sites of cassava and sweet potato varieties, establishing tree nurseries and vaccinating local poultry against Newcastle disease. VBAs sell farmers a range of seed varieties, e.g. new varieties of maize, phaseolus beans, cowpea, cassava, sweet potato, pigeon pea, pumpkin, tomato and lablab. They also sell improved fertilisers in small packs, teach farmers how to make compost, how to space seed and place fertiliser correctly, how to identify and break soil hard pans and how to harvest rainfall.

c) *Business development:* FIPS-Africa encourages its advisers to generate income from selling fertilisers and seeds directly to farmers. Advisers also generate income from vaccinating local poultry. This motivates staff to reach more farmers, and provides an exit strategy for staff after cessation of donor funding. In addition, FIPS-Africa works closely with two firms in the private mining sector to supply small affordable packs of fertiliser, thus creating jobs in the formal sector. The intention is to formalise this relationship so as to enable FIPS-Africa to benefit financially.

RNRRS projects

There are several RNRRS outputs that the FIPS team have made use of. In particular, much use has been made of RNRRS outputs that addressed maize streak virus disease (MSVD) which is a major constraint to maize production in sub-Saharan Africa. Previous research had investigated the role of insect vectors in the epidemiology and ecology of maize streak disease. Efforts were also made to develop management strategies for control of the disease that were acceptable and profitable for East African farmers. One of the main control measures was the production and use of good quality seed from a superior MSV-resistant maize variety. The research made sure that the disease resistant varieties were appropriate to farmer needs, adapted to local conditions and validated by the farmers. The best training methods for using and production of these superior seeds were also tested.

Use has also been made of research that looked at factors affecting the usage patterns of improved weed control practices more generally, which in turn produced lessons learnt about policy measures that affect the uptake of improved weed control practices. Another RNRRS project showed what could be done through a simple small pack marketing approach based on maize and fertiliser by small-scale farmers in Kenya, resulting in a viable integrated maize crop management system. This was followed up by a related project which expanded the variety of inputs available for purchase by farmers and explored the opportunity to take this work forward in Tanzania. All of this work helped to ensure that approaches for improving access to and management of quality seed by farmers was validated and promoted and that sustainable pathways/ systems for quality seed supply appropriate to local conditions and farmers needs were developed by farmers and other stakeholders. In summary, the projects were:

- *R5237 – Biology and management of termites and white grubs in smallholder cropping systems*
- *R6642 – Development of management strategies for maize streak virus disease – 1*
- *R7404 – Socio-economic study of the uptake of herbicide technology in maize cropping systems*
- *R7429 – Development of management strategies for maize streak virus disease – 2*
- *R8219 – Improved access to appropriate farm inputs for integrated maize crop management by small-scale farmers in Embu and Kirinyaga districts, Kenya*
- *R8220 – Improving farmer access to and management of disease resistant cultivars in the southern Highlands of Tanzania*
- *R8455 (2005) – Improved access to appropriate farm inputs for integrated maize crop management by small-scale farmers in Kenya and Tanzania*

Project

The contributing organisations were:

- Farm Input Promotions Africa Ltd., to co-ordinate the initiative, establish and maintain networks of Village-based Agricultural Advisors, access appropriate research outputs from NARCs and private sector companies, and link with Well Told Story Ltd. to design appropriate extension messages.
- University of Exeter, Prof. J. M. Anderson, to advise on dissemination of reduced tillage techniques for improvement of soil health and water harvesting/conservation.
- Dr D. Harris (ex RNRRS/Bangor University) to advise on dissemination of seed priming technology. Dr Harris is now with ICRISAT in Nairobi and has a proposal on seed priming with IFAD. This has been funded with £20,000 each for FIPS and Real IPM.
- Minjingu Mines & Fertilizer Ltd. to provide its new fertiliser blends (*Minjingu Mazao*) to service providers for demonstrations.
- Athi River Mining Ltd. to provide its new fertiliser blends (*Mavuno*) to service providers for demonstrations, and its fertilisers in small packs for sale to farmers.

- Leldet Ltd. to provide seed for demonstrations, arrange its seeds of improved crop varieties in small affordable packages, and distribute its seeds to service providers for sale to farmers.
- Western Seed Co. to do the same.
- Pannar Seed Co. to do the same.
- Osho Chemical Industries Ltd. to supply thermo-stable Newcastle disease vaccine and training in vaccine administration.
- Well Told Story Ltd. to manage the communications elements of the project as a component of its *Shujaaz Youth Communications Initiative*.

The overall project objectives were to further develop and implement an innovative methodology to put research into use to improve the food security of 1,500,000 small-holder farmer families in Kenya, Uganda and Tanzania; and establish, in co-operation with private sector companies, ten networks of 150 village-based agricultural advisors in Kenya, Uganda and Tanzania, to deliver, on a sustainable basis, new fertiliser blends and crop varieties, and information on improved crop and livestock management practices to small-holder farmers. FIPS-Africa had already established that a network of 15 promoters can co-ordinate 15,000 small plot demonstrations on different farmers' fields per year. If ten neighbours benefit from each demonstration, then 150,000 small-scale farmers per year could benefit.

It was therefore proposed that a total of ten networks be employed to benefit a total of 1,500,000 farmer families. Hence, assuming an average of five persons/family, this could impact on the lives of 7,500,000 people by the end of the first phase of the project. FIPS-Africa would work closely with input-supply companies and National Agricultural Research Centres (NARCs) to create demand for new varieties of maize, beans, cowpea, pigeon pea, cassava and sweet potatoes, and improved fertiliser blends. It would also integrate outputs from other RNRRS projects (where appropriate) to improve soil, crop, and livestock management. Links with private sector companies would offer income-generating opportunities for the Agricultural Advisors and ensure sustainable delivery of research outputs by the end of the project.

Progress

Techno-economic

The following progress had been made by June 2011:

- FIPS has reviewed options using pilot studies and trial plots, for bringing soil management and seed priming into its approach; both have been successful although seed priming is most successful in combination with other inputs. Options included conservation tillage and seed priming. More than 78,000 on-farm demonstrations had been completed within the first two seasons (12 months). Demonstrations included better soil management options, use of improved seed varieties, seed priming etc. The impact of FIPS has been increased food security through increased production, faster production, and diversified food sources with reduced risk to the farmers. Support for the FIPS logistics and administration has been accomplished through employing more staff. This has meant that FIPS has become a more efficient organisation, able to operate on a wider scale, accessing more research outputs and with more accessible baseline information and information on its impact.

- As outlined above, the FIPS approach is to use the small pack approach where farmers can experiment with small amounts of new inputs (and improved management), which pose little risk to them. Currently, the FIPS seed repackaging unit repackages 4,000 to 8,000 small packs of seed per day; with five days per week and 240 days per year, they can produce about 1,440,000 small packs per year. A new Farmer-Yield-Share (FYS) seed small pack has been developed, where farmers pay VBAs for the seed by sharing their yield. This has been tested for butternut pumpkin, hybrid tomato and cabbage seeds. In some areas these vegetables performed well and gave substantial food and income to the smallholder farmers from only very small amounts of seed. Butternut squash is particularly exciting because it requires few inputs, stores well and tastes good, providing options for food security. It is also of high value; one pumpkin sells for 0.40 US$ and a 10-seed pack can produce up to 300 pumpkins. The cash benefit from tomatoes can be high but they do not store well and require careful management and spraying against diseases. This is also potentially very lucrative for VBAs who can for example earn up to Ksh 200 ($2) from selling four butternut pumpkins, half of which will be paid to FIPS to cover the cost of seed purchase, packaging and distribution. However, the review showed that collecting the yield was bulky and difficult. In the future, an alternative would be to try collecting the cash value of the yield.
- However, this changed approach can only work in new areas. FIPS argues that farmers are unhappy about change repayment conditions after the start of activity. As a result, district coordinators were not benefiting from the process, so they were not motivated to follow up and push for VBAs to return the crop. FIPS now suggest that a portion of the yield share goes to district coordinators to motivate them to follow up better. District co-ordinators are placing the orders for seed; these include tomato, cabbage, onion, kale, carrot, spinach and coriander. FIPS initially distributed 20,000 packets. Advisors can earn Ksh 3/packet. FIPS also developed the sale of maize seed and ordered 15,000 packets for distribution. The margin for VBAs is KSh 4/packet.

- On economic impact, FIPS has carried out a series of impact assessments (c. 6 as of March 2012). Though these have not yet been fully written up and analysed, summary details are set out in Chapter 7. For example, one recent paper shows the results of 266 household surveys; these were randomly selected and carried out in target areas of ten FIPS-Africa VBAs who had a target of disseminating agricultural technologies (e.g. maize, beans, sweet potato, cassava, chicken husbandry) to 10,000 farmer households in Emuhaya, Sabatia, Vihiga, Hamisi, Kakamega and Butere Districts in Western Kenya. Average farm size was 0.8 acres. In the long rainy season, average maize production increased 2.4 fold from 268 to 641 kg. In the short rainy season, maize production increased 3.2 fold from 118 to 379 kg/household. Around 98 per cent of households reported an increase in maize production. Increases in productivity were attributed to adoption of improved maize varieties and improved crop management practices. Before FIPS, 0 per cent of households reported having enough maize for their families' needs. After three years of FIPS interventions, 95 per cent of farmers reported they had enough maize for their families' needs. Increases in productivity were also recorded for beans, sweet potatoes and indigenous chickens. Data for these technologies are still being analysed. A second paper shows similar data. *These are dramatic figures that indicate very high rates of return.* To carry out surveys, FIPS use field enumerators from local universities (placement students, indicating useful capacity building of local institutions).

Table 6.1 Gender impact

Activity	*Number*	*Proportion of recorded primary beneficiaries female*
Field days	861	52% female
Cassava sales	115,600	80% female
Sweet potato vine sales	269,867	50% female
Cassava bulking sites	8568	61% female
Sweet potato bulking sites	13,988	47% female
Poultry vaccination	106,341 Birds vaccinated	58% female
Poultry dye	7256 Birds vaccinated	55% female

Source: Internal RIU document

- There have been interesting impacts on social inclusion as may be seen from Table 6.1 which shows the results of data collected by the FIPS team in 2010.
- An unintended consequence of self-employed VBAs is that it is more difficult to motivate them (they are often of low education standard) to report on activities carried out. FIPS-Africa has a large range of activities and there is a risk that reporting may become too burdensome.

Capacity

The following progress had been made by June 2011:

- Work on the ground is still mainly in Kenya and Tanzania but has continued to grow rapidly. For example in the six months from June to December 2011, the number of VBAs in place rose from c. 148–c. 300. They covered three regions in Kenya (central, eastern and western) and a number in Tanzania. The operational structure is one in which each region has a regional co-ordinator. In all, there are some 25 districts with each district having some 12 VBAs managed by a district co-ordinator.
- The VBAs were initially trained (by district co-ordinators) in water harvesting, soil management, early maturing varieties, drought tolerant crops and varieties, and diversification of crops away from maize. They then held field days around the priming trial plots to educate farmers and demonstrate the technology. FIPS have explored options for VBA income generation to improve sustainability; for example VBAs have been used on a self-employed basis and have been encouraged to open bank accounts for the first time. This has had the effect of simplifying FIPS-Africa's payment and financial systems. It may also allow VBAs to apply for credit/loans from banks or input suppliers.
- In early 2011, on the advice of RIU, a Business Development Assistant was taken on to assist the VBAs. The introduction of VBAs has been the main method of generating new employment, and FIPS have experimented with various different approaches to make more the number of VBAs by increasing their scope of work and their income.[4] For example, FIPS paid new VBAs an allowance in only 2–4 months of the year to conduct demos, disseminate small packs etc. This reduces potential VBA dependency on monthly allowance, and encourages them to focus more on generating income from offering farmers' services. This also reduces the cost of network establishment, and provides opportunities for more VBAs at the same cost.

Linkages/partnerships

The following progress had been made by June 2011:

- FIPS are now building a strong personal and professional relationship with other Best Bet projects, particularly Real IPM.[5] CIP networks for Irish Potato seed and KMDP networks for maize are benefiting from RIU methods with other crops, livestock and soil/water management. Networks of VBAs funded by CIP for dissemination of Irish potato are also being used to disseminate other crop varieties promoted by the RIU e.g. sweet potato, maize and cowpea, and to vaccinate chickens.
- The FIPS Business Development Executive has and is establishing links with BecA, ILRI, East African Seed, Gatsby, Aquifer Limited, Scan Ad (advertising company in East Africa), AGRA, McKnight Foundation, University of Aberystwyth (Dr D Harris), ACDI/VOCA and AVRDC (Arusha).
- FIPS have also linked with RIU Rwanda who are applying the FIPS approach for stimulating demand for agricultural inputs. Demonstration plots at cell or village level for early maturing maize and two varieties of climbing bean have been established. A system of training VBAs has been established in collaboration with a local seed company (Rwaseco). Vegetable promotion has been conducted in collaboration with Rwanda School Garden Initiative.
- There are possible connections with the Gates Foundation (vegetatively propagated crops) and AGRA (to extend FIPS work into Mozambique, Rwanda and Uganda). Additionally, CropLife International are interested in working with FIPS on the promotion of herbicides.
- Shujaaz – FIPS have received approximately 50 enquiries from farmers via Shujaaz seeking sweet potato varieties which were passed onto FIPS regional co-ordinators.
- Seed companies – to address the shortage of seed, FIPS has sought to strengthen its relationships with more seed companies and to focus more on sweet potato, cassava and poultry. VBA activities have been expanded to include sale of tissue culture banana plantlets, establishment of improved rabbit and pig breeding units, and rearing and sale of improved chicken breeds (layers).
- FIPS are investigating new methods of VBAs' repayment of improved livestock breeds. VBAs are currently being offered loans of a pair of an improved rabbit breed. They have agreed to repay FIPS the first litter which FIPS can then sell or use for capitalising new networks of VBAs. This concept is also being tested for improved pig breeds in the RIU Vihiga network.
- As a result of FIPS activities, Minjingu mines (Tanzania) are now exporting their fertiliser to Kenya. To date they have supplied 11 tonnes of Minjingu granules for sale through its VBAs in Western Kenya.

- The FIPS methodology has now been exported to Nigeria. Here there has been fast growth with some 600 VBAs (promoters) working on contract with a Nigerian seed company funded partially by DFID through its Propcom private sector initiative. This represents a growth from scratch since January 2010. The crops involved are mainly maize and vegetables (c. 3.5 million bags).
- FIPS has current proposals at the moment with USAID ($2m), AGRA ($2m), Rockefeller ($0.5m) and Gates. The USAID proposal has now been funded. There is also renewed interest on the part of the World Bank who invited FIPS to make a presentation at a closed workshop in Washington in May 2012.
- On FIPS methodology, the project has stressed the following characteristics:
  a) FIPS projects are multi-technology and context specific.
  b) They are also small-scale.
  c) They are specifically implemented to minimise farmer risk.
  d) The VBAs are incentivised to ensure success (these incentives are not just monetary but relate also to social capital) and also to provide follow-up services.
  e) In short they are dramatically different from conventional extension modes and represent a key output for the RIU.

Well Told Story Ltd. /Shujaaz

Area of intervention

Well Told Story Ltd. (WTS) was started in 2009 with support from the DFID (RIU) programme that provided 25 per cent of their income until June 2011. However, WTS is an entirely different instrument for getting research into use than the other Best Bets in so far as it uses multimedia to transfer technologies to people through print media and radio. As outlined above Shujaaz was originally partnered by Farm Inputs Promotions Africa (FIPs), when the proposal was presented to the RIU in November 2009 as part of the Best Bets selection process. The plan was for the Shujaaz Youth Communications Initiative (SYCI), developed by WTS to transfer technologies using three forms of powerful and accessible youth focused media:

i) nationally distributed monthly comic books (300,000 every month)
ii) daily syndicated FM radio programmes
iii) interactive SMS to provide youth with ideas for generating income through agricultural technologies.

FIPS would not only be a source of information but also of contacts for the audience to follow up with. The aim was to reach an estimated >6 million youths each month and in doing so to put them directly in touch with FIPS-Africa's extension networks and/or to inspire them to engage with the ideas and technologies promoted. However, it was recommended, at the review stage, that the two elements of this Best Bet, FIPS and Shujaaz, should be funded separately but maintain close links. For the first six months the teams agreed that Shujaaz would source its material from FIPS, after that period it would look for other sources of information. The RIU book of RNRRS outputs was also provided at this time which gave the Shujaaz team additional material to consider. It was anticipated at that stage that 'the SYCI will address the problem of chronic lack of information and lack of engagement of youth in agricultural activities in rural areas'.[6] As of now, Shujaaz uses two main media forms (comics and radio) but is planning to expand into television.

Every Shujaaz comic runs four stories through its four comic characters every month. Each of these stories is usually funded through separate donors or funders. This approach relies on donors or funders supporting general development stories or (as with the RIU) commissioning Shujaaz to deliver particular thematic messages. For example, the NGO Twaweza has supported Shujaaz since this delivers empowerment stories to youth, which aligns with its own mandate. RIU pays for one agricultural story every month. Maintaining a mixed funding base is a strategy to prevent financial dependency on any one donor and maintain editorial independence. However, it has proved hard to coordinate and streamline donor funding in order to ensure that costs are covered for each issue. This has resulted in a variation in the numbers of comics printed and wide differences in the amount of comics passed to M-Pesa agents. Shujaaz also relies on technical inputs in the form of advice from FIPs, Real IPM and other agricultural experts. It has also built up a network of agricultural experts who give *pro bono* advice and support. The agricultural stories covered to April 2011 are listed in Table 6.2 and are a reflection of these technical inputs from FIPs.

RNRRS projects

Shujaaz used the following research outputs in their comics:

- *R5539 – Commercialisation of solar drying technologies for micro and small-scale rural enterprise development*
- *R6619 – Husbandry strategies for forages to increase milk production from cows and goats on smallholder farms in Tanzania*
- *R6762 – Decision tools to aid Armyworm surveillance and outbreak prediction*
- *R7571 – Management of virus diseases of vegetable crops in Kenya*
- *R7966 – Identifying the factors causing outbreaks of Armyworm as part of improved monitoring and forecasting systems*

- *R8312 – Promotion of quality vegetable seed in Kenya*
- *R8407 – Economic evaluation and international implementation of community-based forecasting of Armyworm*
- *R8439 – Promotion of quality kale seed in Kenya*

Project

This Best Bet aimed to communicate entrepreneurial opportunities to young people in east Africa. WTS had been combining the production of a comic book called Shujaaz which is written in the local Kenyan language, with an interactive SMS service by mobile phone and radio broadcasts to help disseminate FIPS work. Comics are a highly accessible communication tool even to those who may not be completely literate, and they are affordable and durable. Comics are extremely popular, with a large percentage of the younger generation reading and learning from them. Approximately 30 per cent of Kenyans under 35 had read a WTS comic book in 2010.

In order to make the comics even more effective the intervention was combined with messages on the radio and feedback from SMS text messages. The plan was for the Shujaaz Youth Communications Initiative (SYCI) developed by Well Told Story Ltd to use three powerful and accessible youth-focused media: nationally distributed monthly comic books (300,000 every month), daily syndicated (five minute) FM radio programmes and interactive SMS to provide youth with ideas for generating income through agricultural activities. Listening to the radio is still something that the majority of people in Africa do; in Kenya more than 90 per cent of people listen to FM radio every day and so it was seen as an excellent vehicle for promoting FIPS/RIU messages. The more recent success of mobile phone technology is also well known; mobile subscriptions in Africa rose from 54m to almost 350m between 2003 and 2008, the quickest growth in the world. Mobile phone growth in Kenya has been remarkable, even among the rural poor. In June 1999, Kenya had 15,000 mobile subscribers. Today it has nearly 8 million out of a population of 35 million, and the current operator networks are as extensive as the access to banks is limited. It is true that the poor often cannot afford to have a mobile phone subscription, but in Uganda for example street vendors offer mobile access on a per-call basis. They also invite those without access to electricity to charge their phones using car batteries.

The aims of the project were the production of a monthly comic book promoting agricultural research which farmers can adopt along with additional material, a website, facebook, radio programme and Twitter feed. WTS would also construct mechanisms for readers to provide feedback and enter into dialogue with the research expertise. Comic books would be distributed through the *Daily Nation* newspapers and Safaricom

M-Pesa kiosks, accompanying daily radio programmes broadcast on a national syndication network of partner FM radio stations. The team would produce storylines with a clear call to action, combined with the information necessary for audiences to take action from which feedback would be encouraged. Audience feedback would be shared with the information providers so that further follow up could take place.

Progress

Techno-economic

Table 6.2 lists the production and circulation of comics and their radio impact over the period March 2010 to April 2011. According to WTS by the end of 2011, 36 per cent of respondents said they had taken action based on the Shujaaz stories and a further 32 per cent that they had spoken to others about ideas from Shujaaz. Copies of Shujaaz comics carrying RIU stories and a letter were sent on two separate occasions directly to more than 100 targeted Kenya MPs and policy-makers. Preliminary impact research on this activity indicates that significant attention and influence has been achieved by this means.

Table 6.2 Comic stories 2010/2011

Chapta	*Month*	*RIU Story*	*Circulation*	*Other media*
1	March	Dyeing your chickens to evade predatory birds	200,000 in Saturday Nation 300,000* through 9,000 M-Pesa agents**	2 radio stations QFM and Easy FM
2	April	Breeding termites for chicken feed	200,000 in Saturday Nation 100,000 through M-Pesa	5 radio stations
3	May	Box bailing of maize stover	200,000 in Saturday Nation 100,000 through M-Pesa	
4	June	Solar drying of fruit AND improved sweet potato	200,000 in Saturday Nation 400,000 through M-Pesa***	14 radio stations (6 in the remote north) 8,764 hits on website Wide-scale use of Facebook page

Table 6.2 continued

Chapta	*Month*	*RIU Story*	*Circulation*	*Other media*
5	July	Chicken vaccines	200,000 in Saturday Nation 400,000 through M-Pesa	14 radio stations 12,547 hits on website Wide-scale use of Facebook page
6	August	Seed priming	200,000 in Saturday Nation 400,000 through M-Pesa	14 radio stations 14,529 hits on website Wide-scale use of Facebook page
7	Sept	Armyworm forecasting	220,000 in Saturday Nation, 280,000 through M-Pesa (Nation increased sales)	18 radio stations over 20,000 hits on website Wide-scale use of Facebook and Twitter
8	Oct	Healthy seed	220,000 in Saturday Nation, 180,000 through M-Pesa	18 radio stations Wide-scale use of Facebook page
9	Nov	New varieties of maize	220,000 in Saturday Nation, 80,000 through M-Pesa	18 radio stations Use of Facebook and Twitter
10	Dec	Stored seed	220,000 in Saturday Nation, 80,000 through M-Pesa	18 radio stations Use of Facebook and Twitter
11	Jan	Chicks in a pen	220,000 in Saturday Nation, 80,000 through M-Pesa	20 radio stations
12	Feb	Recaps of key RIU stories [anniversary special]	We hope to resume 600,000 from February – depending on income received	20 radio stations
13	March	Fish Farming [TBC]	Depending on continuing RIU funds	
14	April	Fish drying [TBC]	Depending on continuing RIU funds	

Source: Internal RIU document

At least 10 million Kenyans have been exposed to Shujaaz and more than 1.52 million Kenyans can be described as Shujaaz 'core audience' who are known to have followed Shujaaz closely and discussed and applied ideas and innovations. New media – i.e. Facebook, SMS etc – accounted for more than 250,000 audience conversations during the first seven months of 2011, and close to 100,000 SMS text messages. The audience was split 60/40 per cent male/female. Facebook, Twitter and SMS were also established as active channels for audience interaction with Shujaaz and its content. More than 10,000 Kenyans follow Shujaaz on Facebook. Twitter is rapidly emerging as a new medium for audience interaction, gaining more followers daily – currently 1,000. WTS have not been able to disaggregate this audience by gender, since this information is not disclosed by users. There were 8.5 million comic books containing specific guidance on RIU-informed innovations published and distributed nationally. Feedback evidence confirms that these are still in circulation and being read.

In September 2011 GTZ, in collaboration with Well Told Stories and the National Commission on Cohesion and Integration (NCIC) of Kenya, commissioned Synovate to carry out an audience survey of Shujaaz. They found that Shujaaz is now Kenya's largest single independent communications channel. Fifteen million Kenyans under 35 say they are aware of Shujaaz, the bulk of the readers are between ten and 33 years old, with a peak at 18–26 years; 36 per cent of audiences say they have acted directly on information in Shujaaz (Nijihia, 2010). A qualitative component comprising of ethnographic case studies was conducted among eight pre-selected readers of the comic who indicated that the comic had impacted their life in some way. The objective of the ethnographies was to understand, in depth, the lives as well as the views, experiences and relationships of readers who had adopted either a practical innovation or a peace-building initiative as a result of reading the comic. Data were collected using observation, structured, un-structured and semi-structured guides, and a listing of members of the readers' social networks.

Capacity

WTS is seeking advice and is negotiating with Twaweza and other donors, as well as with the partners through whom they distribute the comic, and with other corporate bodies, so as to secure longer term commitment. Furthermore, there is recognition that the WTS team have skills which could be used to generate income, for example through being contracted by NGOs and regional/continental bodies (such as AGRA and ASARECA) to prepare commissioned multi-media products focusing on agriculture and other development-related topics. With regard to scalability, Shujaaz already has a potential for outreach including to other countries in the region. Indeed RIU always felt this to be a reasonable aspiration.

Linkages/partnerships

Extensive partnerships have been established between Shujaaz/Well Told Story and the private and public sector in Kenya (including with Google, Nokia, USAID, GTZ, etc). WTS also believe that they have had a direct positive influence on the research teams we have worked with, many of whom have come to see new possibilities and potency of communications in their work. An independent review by the IDL group (IDL, 2011) considered that Shujaaz is innovative in several ways. With regard to putting research into use, WTS believe Shujaaz is an effective and innovative means to communicate with large numbers of male and female youth, in a contemporary and 'fun' language that they relate to. It has the power to change the way youth see agriculture and to broaden livelihood options for them accordingly.

In addition, the way in which WTS has used far reaching distribution mechanisms – the *Nation* newspaper, and the M-Pesa money transfer outlets across the country – is in itself innovative. Twaweza, which specialises in funding work through large networks and institutions – such as mass media, mobile phones, religion and consumer goods networks – commented that of all the programmes they support, Shujaaz is the most innovative in that it is using several media (radio, Facebook, comics, SMS). This means of distribution allows for Shujaaz to have impact and influence at scale. Institutional impact of this experiment in the sense of influencing relationships between key actors has been limited. Donor funding from RIU, GTZ and Twaweza has all been on a relatively short-term basis to date and this has left WTS in an uncomfortable situation financially. On the positive side, the Shujaaz Best Bet won the International One World Media Special Award in May 2011 and most recently (April 2012) was awarded the 2012 International Digital Emmy Award for best programme in the Children and Young People category. This has facilitated special internal mention from DFID and from the UK Minister for International Development on a phone-in at Shujaaz FM.

Stamping out Sleeping Sickness (SOS)

Area of intervention

Sleeping sickness is used to describe two quite distinct diseases caused by different sub-species of *Trypanosoma brucei* – *Trypanosoma brucei gambiense* and *T. brucei rhodesiense* are both human infective; the third sub-species of *Trypanosoma brucei*, *T. brucei brucei*, is morphologically indistinguishable from the others and infects a range of mammalian species, both domestic and wild, but is not human infective. *T. b. gambiense* and *T. b. rhodesiense* are invariably fatal in humans if left untreated. These two diseases have distinct clinical pictures and their discrete geographical distribution across Africa serves to emphasise these differences; understanding these biological, clinical and epidemiological differences has been crucial in devising effective methods of disease control and in defining health policy in relation to sleeping sickness (Maudlin, 2006). One of the main factors stimulating current remedial measures is the fear that spread of these two strains is threatening poor populations in Uganda (Welburn *et al.*, 2006). Cross impact of the two parasites could lead to major problems in diagnosis and treatment of the disease. The Stamp out Sleeping Sickness (SOS) project has now been operational for some six years.

RNRRS projects

There have been a whole series of these managed by the DFID CRD Animal Health Programme. They deal with the parasite itself, the vector (tsetse fly), appropriate diagnostic and treatment regimes and socioeconomic studies of impact and output. In summary, these are as listed below.

- *R7364 – Improving the control of tsetse: the use of DNA profiling to establish the feeding of tsetse to cattle (April 1999–March 2001)*
- *R7360 – Field methods and tools for resource-poor farmers and extension workers to improve targeting and appropriate use of drugs used to control African bovine trypanosomiasis (April 1999–March 2005)*
- *R7538 – Review of environmental change and sustainable, poverty-focused strategies for trypanosomiasis control in Africa (December 1999–June 2000)*
- *R7596 – Decision support system for the control of trypanosomiasis in south-east Uganda: improving public health and livestock productivity through cost-effective control of trypanosomiasis in livestock (April 2000–March 2006)*
- *R7597 – Development of a low-cost haemoglobinometer and other diagnostic tools for bovine disease diagnosis in sub-Saharan Africa (April 2000–March 2003)*
- *R7539 – Environmental risks of insecticide-treated cattle in semi-arid Africa livestock systems (January 2000–August 2002)*
- *R7987 – Message in a bottle: disseminating tsetse control technologies (October 2001–September 2004)*
- *R8214 – Integrated vector management: controlling malaria and trypanosomiasis with insecticide-treated cattle (October 2002–March 2006)*
- *R8318 – Decision-support for endemic diseases in sub-Saharan Africa, private sector drivers for technology adoption by poor livestock keepers (September 2003–March 2006)*

- *R8461 – General model for ITC restricted application of insecticide to cattle (September 2003–December 2004)*
- *R8459 – Tsetse Muse: an interactive programme to assess the impact of control operations on tsetse populations (February 2005–October 2005)*

Project

SOS is run through a loose consortium based at the Faculty of Veterinary Medicine (FVM) at Makerere University, Uganda. Stakeholder interests include cattle owners, district veterinary officers, universities (mainly Edinburgh and Makerere), the corporate sector (an international drug company and a venture capital firm), local private sector companies, donors and national regulatory authorities. The component being funded by DFID through RIU is an attempt to develop long-term sustainability through a public/private partnership designed to create small veterinary businesses (3V Vets); the belief is that unless steps are taken in this direction veterinary care will continue to remain within the province of charities such as NGOs and the corporate social responsibility (CSR) activities of international private companies. As such it will remain a dependent activity reliant on outside inputs and in this sense not integrated into national capacity building. The challenge is really one of how to bring this new sustainability about. The Best Bet was therefore to support Phase 2 of SOS activity. It began operations in January 2010.

Progress

Techno-economic

As of December 2011, considerable progress has been made. RAP (restricted application of insecticide technology developed under RNRRS research at NRI) and veterinary services are now available in seven of the 18 SS affected districts of Uganda that are home to some 500,000 cattle. Around 100,000 animals are being regularly sprayed which is sufficient to control both animal and human trypanosomiasis (HAT) at village level. Farmers report animals treated are healthier, more productive and fed better. They are also protected against a range of other tick-borne diseases such as theileriosis and cowdriosis. Table 6.3 illustrates SOS3V provision of veterinary care established in seven districts to date: Dokolo, Kabaramiado, Amolitar, Serere, Soroti, Lira and Apac.

Tororo will soon be trialled to validate the percentage cover needed in the area and one further district will be included after assignment by the Uganda government. This equates to 74,000 head of cattle across seven districts under SOS3V vet service provision out of a total cattle population of 1,070,518 equalling 7 per cent. It is estimated that an equal number of cattle are being sprayed using other products in the system that are active against both tsetse and ticks but that are being used as dip formulation or 'pour on'. Lira and Apac are using a product called tsetse tick – marketed locally – that is active against tsetse and ticks and it has been more difficult for SOS3V to gain a foothold in this market (while subsidised products via NGOs are in the system).

From the human health perspective, numbers of new cases have been consistently falling in the SOS area from over 500 prior to inception of SOS to below 125. While this is welcome news the parallel RIU health study has clearly shown that when disease is moved into a new area from cattle migration, underreporting of disease can be extremely high and mini epidemic situations arise with infection rates in human population of over one per cent in particular affected districts. Medical services are estimated to be finding less than ten per cent of cases using the present means of surveillance. A new method of screening for infection is highly specific and sensitive at capturing early stage (more treatable cases) and a new method of surveillance is under discussion taking into account new risk zones (associated with establishment of new markets as the principal risk factor).

Indeed a modelling study using parameters for Uganda has calculated that relatively modest levels of treatment (~20 per cent of animals, even if tsetse numbers are not reduced by the intervention) could lead to the elimination of HAT in SE Uganda. The work commissioned by the study

Table 6.3 Veterinary care provision by district

District	*Veterinary Care Provision*
Dokolo:	cattle population 54,336 of which 12,000 sprayed per month by SOS3V service provision = 22% under treatment
Kaberamaido:	cattle population 87,569 of which 5,000 sprayed per month by SOS3V service provision= 5.7% under treatment
Amolatar:	cattle population 122,739 of which 12,000 sprayed per month by SOS3V service provision= 9.78% under treatment
Lira:	cattle population 127,833 of which 6,000 sprayed per month by SOS3V service provision = 4.69% under treatment
Apac:	cattle population 241,269 of which 5,000 sprayed per month by SOS3V service provision = 2.07% under treatment
Soroti:	cattle population 356,777 of which 18,000 sprayed per month by SOS3V service provision = 5% under treatment
Serere:	cattle population 190,461 of which 16,000 sprayed per month by SOS3V service provision = 8.4% under treatment

Source: Internal RIU document

has now been published as an RIU output in PLOS NTD, the highest impact journal in this field (Hargrove *et al.*, 2012). The paper shows that insecticide-treated cattle are always going to be a better bet than trypanocides and that treating a reasonable proportion of cattle with insecticides can lead to total eradication of the disease. Dokolo, Kabaramiado, Amolitar and Serere are high HAT risk areas but have medium tsetse density and farmers are treating >15 per cent animals per month. Currently SOS 3V has a 50 per cent share in this market, which is rising. In Soroti there is a low density of tsetse and farmers are treating >15 per cent animals per month. Lira and Apac are HAT low risk but have a greater fly presence – farmers are treating in excess of 15 per cent cattle per month but SOS 3V have a lower percentage of market share in these districts. Hence, disease is being controlled in all areas; there are some isolated outbreaks near markets from imports of new 'dirty animals' but coverage is containing spread.[7]

Capacity

As one component, the Faculty of Veterinary Medicine created a new *Institute for Strategic Animal Resource Services (AFRISA)* linked to (but financially independent of) the University of Makerere.[8] Part of this new institute programme is a body designed for in-training community service delivery. The University sees this as a generic mechanism for equipping graduates for a labour market that is no longer satisfied by the supply of traditional university degree-holders. Instead the demand is for graduates that not only possess saleable business skills but are also capable of actually generating their own jobs virtually from scratch. Under this programme veterinary students spend the final year of an undergraduate degree entirely in economic production activity producing at the end a project report that is assessed as a key component of the final degree. In the SOS case and in co-operation with a private veterinary company, final year undergraduates participate in block treatment of cattle and ancillary spraying activities.

In addition, a small number of these undergraduates have been encouraged to set up as small 'agrovet' businesses (3V Vets) under the supervision of the private veterinary company. Undergraduate vets are trained in community animal health services and gain three months 'short course' practical experience. To date seven businesses are now established with 100 additional employment opportunities created – each vet has a shop assistant and between 90–100 spray persons are employed to date by these seven vets.[9] Aside from RAP, farmers are buying drugs for helminths, *trypanosomiasis* and tick-borne diseases from the 3V network of vets and a PPP vet service is now available in all districts. Twenty-seven BVM students were trained in Phase 2 in Soroti. The AFRISA approach is now being examined by other African universities and cognate work has now begun in Nigeria. There is also one SOS 3V manager and assistant and the AFRISA office comprising a director, manager and two assistants. The presence of vets providing service is universally welcomed across the districts.

Linkages

There are a wide range of these. Readers are referred to Morton (2010) for further details.

Broad summary conclusions

This chapter has summarised the remaining three Best Bet projects. Two of them (FIPS and SOS), represent extensions to existing on-going projects.[10] They were supported by RIU because they met the selection criteria specified in the original call for concept notes and had performed well in the selection process. In both cases there were strong private sector potential components involved. FIPS has clearly performed well in all respects. In two years, it has expanded its operations rapidly, put many RNRRS outputs directly into use, publicised a technology development model that has impacted influential stakeholder groups such as the World Bank, mobilised and extended private sector resources, had considerable influence on targeted smallholder farming communities, and demonstrated considerable uptake. It has also generated interest at venture capital level promoting discussions that are on-going.

The third (WTS/Shujaaz) was a new venture but had been included originally as part of the FIPS bid but was then separated since it was rather different from the other Best Bets in a number of respects. The primary difference was that rather than being interventionist in a directly productive sense, its likely impact lay in its potential for enfranchising unemployed youth. The hope was that by showing ways in which simple technologies could be used productively it would catalyse entrepreneurial activity on the part of sections of LIC communities that had limited exposure to traditional types of development aid. It is finally worth noting the cross-interaction between Shujaaz and FIPS which illustrates a feature of the whole Best Bets programme, and the use made of many RNRRS project outputs.

The SOS project has also been successful in many respects. Not only has there been productive use made of RNRRS project outputs at the level of disease spread and community health but it has broken ground in terms of building local university resources and how they too might be better organised in the service of economic production. As with FIPS, there are now on-going discussions with venture capital firms. Some further discussion of this point will take place in Chapter 7.

Notes

1 There were some doubts on the part of the Dragons about the viability of Shujaaz, so when the contracts were issued, separate contracts were awarded to FIPS and WTS in order to give Shujaaz a much broader mandate, i.e. not just restricted to FIPS. Over the course of time this has been proved correct.
2 See www.fipsafrica.org/index.php for more information on this case.
3 And their educational level varies widely. What is important to FIPS is the range of 'people skills' needed to interact and build trust with local communities.
4 In fact the rapid expansion of FIPS activity has led to capacity problems; they are actively trying to employ at least five more staff and as of March 2012 a communication person is now in place.
5 See Chapter 4. Real IPM is investigating how to work with locally available and affordable phosphate in a practical manner in Africa, as the variant that works in Asia is not widely available in rural areas of Africa, and is expensive.
6 Internal RIU document.
7 Recent data comparing Uganda and Nigeria may be found in Okello (2013).
8 Outlined in more detail in Chapter 7.
9 By way of an example, one vet is selling in excess of US $2,500 in vet products per month in Serere District and expanding her business.
10 In fact since it was already underway SOS Phase 2 did not go through the 'dragons den' process but was included because it fitted the specified criteria well.

7 Putting research into use

RIU Best Bet programme

Introduction

This monograph has been about putting research into meaningful use. As outlined in Chapter 3, the DFID RNRRS global 11-year Programme run by its Central Research Department (CRD) had come to an end in 2006. At some considerable expense (around £220m) it had funded some 1,600 individual research projects covering most aspects of natural resource management affecting smallholder agriculture in low income countries (LICs). An RNRRS evaluation had also informed the development of DFID's new Strategy for Research on Sustainable Agriculture (SRSA).[1] In short, the CRD determined that research targeted at development was necessary but not sufficient – what was required was additional effort aimed at putting the knowledge into use. Accordingly, it launched a call for consortia whose remit was precisely to fill this gap. Hence within its proposed design for SRSA, CRD included £37.5m for a programme to get more research into use. It was 'designed to validate and promote the best innovations from previous research funded by DFID through the Renewable Natural Resources Research Strategy (RNRRS)'. And it was meant to 'follow on from existing research programmes using similar management arrangements and utilising established institutional networks including governments, NGOs, national research systems, universities, international centres, donors and farmers'. It was also meant to work closely with the sub-regional programmes such as the CAADP.

In this concluding chapter we focus on what the experience of the Best Bet initiative of the RIU can tell us about putting research into use and what policy implications this has for future technology aid to LICs. In the first section we focus directly on the original RNRRS projects. Even at RIU inception only a small minority of these were deemed relevant by RIU management but experience showed that even this was a gross overestimate. This was an important factor leading to the launching of the Best Bet programme as one that could potentially act as a vehicle for more effective technology development. The results appear to show that RNRRS technology has been useful, but not perhaps as was originally intended. Instead of playing a 'supply push' role, the projects (and their scientists and research bodies) seem rather to have played a key interactive role, in a sense facilitating the productive interaction of supply and demand. We then explore production and business issues showing how in some cases there have been positive results while in other cases not. Special attention is paid to how RIU interventions promoted the development of private sector resources targeted at the rural poor and what the obstacles were in these respects.

The next section deals more specifically with contextual factors such as national government regulation; it explores how such factors influenced the ease with which RNRRS project outputs could be put into use. It also deals with capacity building and other forms of institutional change. While not initially considered as an expected output the RIU has shown how its interventions helped to mobilise a whole range of contextual capacities including the tertiary education organisations which will hopefully act as the seed corn of future technology development. The final section brings things together within a framework for future technology development aid. As DFID had made clear at the outset, the RIU was intended as an innovative learning programme designed to suggest a partial alternative to conventional modes of operation. Inevitably mistakes were made and some interventions did not work. But some did. The issue now for DFID is how to build on these RIU lessons for its future strategy.

RNRRS projects

The RNRRS had run from 1995 to 2006 with its aim to 'remove researchable constraints to the sustainable development and/or management of natural resources'. The strategy was managed through ten research programmes: Animal Health, Aquaculture and Fish Genetic Research, Crop Post-harvest, Crop Protection, Fisheries Management Science, Forestry Research, Livestock Production, Natural Resources Systems, Plant Sciences and Post-harvest Fisheries Programme. At RIU inception, of the 1,600 individual research projects, 280 were considered by the ten programme managers to have the highest potential for impact and these were included in the RIU's Natural Resources Knowledge Database.[2]

The ten individual RNRRS research programmes were designed to generate new knowledge and promote its uptake and application. They addressed the needs of people operating in a range of environments who were dependent upon crops, livestock, fisheries or forests for their livelihoods. The results achieved by the programmes and projects implemented under RNRRS showed that while there is potential for agricultural and natural resources research to reduce poverty, promote economic growth and mitigate environmental problems – and thus contribute to the Millennium Development Goals – much of that potential remains unrealised. This is in part because of the difficulties of scaling up the results of research, i.e. multiplying them on a large scale. RIU was therefore originally conceived as an activity that would link together the many agents involved in innovation – policymakers, researchers, suppliers and end users – and enable a system which uses research to benefit the poor, leveraging greater impact from the RNRRS investment and exploring how the lessons learnt could be applied to other research outputs.

An important assumption and rationale behind the funding of RIU was that the RNRRS portfolio of research outputs would provide a rich source of neatly packaged technologies that could be put into use. This proved to be largely unfounded. While some research outputs from the RNRRS programmes have been successfully applied for the benefit of the poor, of the 280 research outputs assembled by RIU in its database, few could be regarded as being amenable to significant scaling up and out. The outcomes were a good deal less than impressive. An internal report on Best Bet possibilities explored the proposition that there were indeed 'low hanging fruit' awaiting exploitation. However, even after excluding unlikely candidates and reducing the number from 280 to 174 it concluded, perhaps naively as we shall see, that 'none of the (remaining) 124 outputs examined should be seen as an obvious best bet or "low hanging fruit"'.

Nevertheless, we have seen in Chapters 4–6 that use has been made of some 61 RNRRS projects, and often in more than one Best Bet.[3] So the question remains, what is it about these funded RIU projects that lets this occur? We believe the answer is actually fairly straightforward. Individual research projects by themselves can only provide part of the knowledge needed for technology development. The entrepreneur is dealing with a total technology system where many elements are needed for successful innovation. These may be financial, managerial or other context specific information where all are necessary at different temporal stages to suit specific circumstances. Very little of this is known *ex ante* but occurs as an essential part of development.

For example, in the Real IPM Gro-plus case, seed priming technology coming from the University of Bangor had been well understood by Dr Harris and his colleagues but needed contextual adaptation for its application. This was geographical (West Kenya), product specific (mainly maize), technical (there was a need for mixing with fertiliser) and financial (the fertiliser mix was too costly and needed adjustment). Furthermore, it subsequently became clear (from the University of Wageningen and comments made by farmers) that the use of Gro-Plus even without the myco herbicide actually helps to reduce striga infestation. Real IPM is now exploring this with a small group of 100 farmers looking specifically at the effect of Gro-Plus on striga infestation. This to some extent will counteract the problems associated with the original Best Bet.

A second example is in the Armyworm project, which has been assisted considerably by continuing UK Research Council support from Lancaster University (Prof K Wilson) and NRI (Dr D Grzywacz). Dr Grzywacz helped draft the production and applications manual for the Arusha facility and Prof Wilson has developed a database to help target high risk areas. The database collates more than 25 years of information from Armyworm outbreaks and from moth traps throughout Tanzania. These data are available in a number of user-friendly tables, graphs and maps, and can be used by the MoAFS national Armyworm forecaster and staff, as well as by Eco Agri Consult Ltd., to help identify areas at greatest risk of Armyworm attack.

Conversely, on a slightly more negative note, the Aquashops Best Bet may have suffered from lack of on-going technical input. The original RNRRS body (Stirling University) withdrew its input at the end of a delayed Phase 1 (June 2011) and it became clear that there have been a range of technical issues still unresolved. These relate to fingerling quality since the biology is complex needing further research, probably over-specified feed quality since artisanal fish farming can be done with fish feeding on natural organisms in ponds, and technical issues on how best feed should be introduced into ponds and on 'one sex' fish farming.

It is also clear that treating the RNRRS projects as one-off occurrences is a gross oversimplification of the research process. For example, the technological history on Armyworm control goes back to 1965 (funded by ODA) and there has been continuous work on it since that point. Similarly research on tsetse and sleeping sickness control has been supported by DFID/ODA over many decades.[4] Hence, while the restricted application of insecticide technology (RAP) and the molecular tool for identifying human infective trypanosomes were developed under RNRRS and may have been the key innovative findings for the current SOS project, it has built on many years of complementary science (see Table 7.1). Finally, it should be noted that in many cases even at a scientific level, RNRRS knowledge has worked alongside knowledge outputs from other research bodies, for example with the IITA input into the seed yam project and the University of Hohenheim and KARI with Stopstriga and BCA. We shall return to this systemic take on RIU innovation at the end of the chapter.

Table 7.1 Past DFID/ODA funding for Tsetse and Trypanosomiasis (1980–2006)

Subject	*Number of projects*	*Total expenditure £'000*	*% of expenditure*
Trypanosome biology	11	2,007	5.3
Tsetse biology	28	4,979	13.1
Sleeping sickness	4	1,101	2.9
Livestock control	15	3,752	9.9
Control measures support	13	2,660	7.0
Geographical info systems	11	1,733	4.6
Small ruminants, trypanosomiasis and trypanotolerance	11	840	2.2
Tsetse control projects	10	7,757	20.5
Institutional support	11	13,092	34.5
Total	114	37,921	100.0

Source: Internal RIU document

Impact and outcomes

Chapters 4, 5 and 6 have shown quite a wide variation in project impact and output among the Best Bets. For those that have been slow getting off the ground (seed yam, Armyworm, Aquashops, BCAs) factors such as weather conditions, unforeseen technological constraints and regulatory requirements have delayed progress though ultimately all show some degree of longer term promise. In the case of Shujaaz, the very nature of this project militates against specific results though there is indirect evidence of impact. Real IPM/Stopstriga has shown positive results if not quite in the way initially envisaged. In the case of NERICA rice there has been partial success, mainly at out-grower level though this has not yet fully developed into a viable commercial set of activities. However, the impact of the SOS and FIPS projects has clearly been considerable and we shall see below that each of these have sparked commercial interest. In the SOS, the block treatment of cattle has effectively created a new market for drugs and small-scale agrovet business while in the FIPS case the expansion of the small-scale VBA system has been rapid.

Although the FIPS RIU programme was outlined in Chapter 6, it is worth exploring the FIPS case in a little more detail since more recent work has been done on evaluating impacts.[5] FIPS have now carried out six preliminary impact assessments as outlined below. They refer mainly to 2011 experience and despite the limitations of this snapshot they provide a very useful insight. In each case enumerators recruited from local universities were directed towards the Village Based Advisors (VBAs) by the district coordinator. The enumerators then randomly selected households within the VBAs' target villages. The broad results are as follows.

Nzaui district of Kenya's semi-arid Eastern province: increased chicken production

Results show that all surveyed households had begun vaccinating chickens and protecting them from predators, resulting in a major decrease in mortality from disease and predators. As a result, the number of chicks per clutch that reached adulthood increased 2.75 fold from four chicks per hen per cycle before FIPS to 11 chicks per hen per cycle at the time of interview. Once chick mortality had been brought under control, flock sizes were able to increase with the average number of reported adult birds increasing by 2.5 fold from 11 birds before FIPS to 27 birds at the time of interview. The number of birds eaten during the past year had doubled from seven birds to 15 birds. The number of birds sold during the past year had doubled from seven birds to 15 birds. The average combined benefit in birds (calculated as current flock size plus birds eaten plus birds sold) increased by 32 birds from 25 to 57 birds with an estimated value of $129. If each VBA provided chicken services to only 250 households and birds sell at KSh 350, this would be worth $98,800 in sales or $395,000 in value to the community. In addition to increased numbers of chickens, households also experienced 2.6 fold increases in the number of eggs for sale and consumption.

Kenya's Keumbu District: increased maize and livestock productivity

After marketing the improved maize varieties in Keumbu district for six seasons, 159 random household surveys were carried out in target villages of two Village-Based Advisors (VBAs). The average farm size was 1.3 acres and the proportion of surveyed households using improved varieties promoted by FIPS-Africa had increased from 78 per cent to 99 per cent. The average maize production increased by 2.3 fold during the long rains (from 319 kg to 704 kg) and 2.2 fold during the short rains (from 314 kg to 717 kg). If these results are representative across the 2,000 households targeted by the VBAs, then the increased production would be valued at $508,000 per year. The increase in production was attributed to improved varieties and management. Farmers also stated that they can now produce enough to feed their families with a 95 per cent increase in the number of households that produce enough food. Before FIPS, only two per cent produced enough. After FIPS intervention, 97 per cent are able to produce enough. Increases in productivity were also recorded for indigenous chickens. Among the 38 surveyed households, 20 per cent increased their flock sizes. The average number of adult chickens increased fourfold and the average number of chickens sold increased sevenfold.

Kenya's Gucha district: increased maize productivity

After marketing the improved maize varieties in Gucha district for six seasons, 184 household surveys were carried out in target villages of two Village-Based Advisors (VBAs). The average farm size was 1.84 ha. and the proportion of surveyed households using improved varieties promoted by FIPS-Africa had increased from 47 per cent to 100 per cent. Farmers reported that the production of the improved maize varieties increased by 1.7 fold during the long rains (from 380 kg to 631 kg) and 1.8 fold during the short rains (from 286 kg to 511 kg). Farmers also stated that they can now produce enough to feed their families with an 81 per cent increase in the proportion of households that produce enough food. Before FIPS, only 17 per cent produced enough; after FIPS intervention 98 per cent are able to produce enough.

Greater Vihiga district in Kenya's western province: improved food security after adopting new sweet potato varieties

In the greater Vihiga region of Western province (Greater Vihiga, Vihiga, Sabatia, Emuhaya and Hamisi), FIPS-Africa's 15 VBAs have been marketing a range of product enhancing technologies to small-holder farmers for drought tolerant crops among approximately 15,000 households. With regard to sweet potato, FIPS-Africa's interventions focused around provision of improved sweet potato varieties combined with advice on general planting, spacing and propagation methodologies among approximately 7,500 households. During the months of April to July 2011, after one and a half years of promotion, 524 household surveys were carried out in randomly selected households within the target district villages to determine the impact on farmers within the promotion zone. Since April/July 2011, FIPS-Africa VBAs have continued to promote the improved varieties to an additional 7,500 estimated households.

Results show that 0.4 per cent of the surveyed households had used the improved sweet potato varieties before FIPS-Africa came to their villages. At the time of survey, approximately 80 per cent of randomly sampled households were using the improved varieties and took part in the survey. Of the surveyed households 67 per cent were using two or more improved varieties of sweet potato. The improved varieties provided food for farmers in a shorter time (6.8 to 3.7 months) and increased production, due to increases in tuber size, area planted and numbers of tubers produced per vine. The total production calculated among the 524 surveyed households increased by an estimated 8,028 tonnes which has a value of approximately KSh 200,700,000 or $2.36 m. If this is consistent across the 7,500 households to have received the improved varieties in the promotion zone at the time of interview, it would have a value of US$ 33.8 million. With regards to food security, the proportion of surveyed households using the improved varieties that had sufficient sweet potato for their families needs increased from 0 per cent before FIPS to 99 per cent at the time of interview. The proportion of surveyed households who attributed their increased sweet potato production to FIPS Africa was 99 per cent.

Siaya district in Kenya's Nyanza Province: improved food security through adoption of new sweet potato varieties

In Siaya district of Nyanza province, FIPS-Africa's eight village-based advisors have been promoting a range of product enhancing technologies to small-holder farmers for drought tolerant crops among approximately 4,000 households. With regard to sweet potato, FIPS-Africa's interventions focused around provision of improved sweet potato vines combined with advice on general planting, spacing and propagation methodologies among approximately 4,000 households. In March 2011, after one year of promotion, 148 household surveys were carried out in randomly selected households within the target villages to determine the impact on farmers within the promotion zone. Since March 2011, FIPS Africa VBAs have continued to promote the improved varieties to an additional 4,000 estimated households.

Results show that none of the surveyed households had used the improved sweet potato varieties before a FIPS-Africa VBA was established in their villages. At the time of survey, 70 per cent of randomly sampled households were using the improved varieties and took part in the survey. Of the surveyed households, 82 per cent were using two or more improved varieties of sweet potato. The improved varieties provided food for farmers in a shorter time (6.7 to 3.6 months) and increased production, primarily due to increases in numbers of tubers produced per vine. The total production calculated among the surveyed households increased by an estimated 5,362 tonnes which has a value of approximately KSh 134,050,000 across the 148 households or US$ 45m across the 4,000 households to have received the improved varieties in the promotion zone at the time of interview. With regards to food security, the proportion of households using the improved varieties that had sufficient sweet potato for their families needs increased from 0 per cent before FIPS to 100 per cent at the time of interview. All of the households attributed their increased sweet potato production to FIPS-Africa.

Kenya's Kiambu District: increased maize productivity

After marketing the improved maize varieties in Kiambu district for two seasons, 140 random household surveys were carried out in target villages of 13 Village-Based Advisors (VBAs). The average farm size was 1.5 acres. Within the target villages, 74 per cent of randomly sampled households reported an increase in maize production. Farmers reported that the average maize production increased by 1.9 fold during the long rains (from 124 kg to 236 kg) and 1.7 fold during the short rains (from 99 kg to 169 kg). This correlated with an average increase in consumption of grain from 4.4 kg to 7.5 kg per household per week over the same period. If these results are representative across the 2,000 households targeted by the VBAs, then the increased production would be valued at US$ 240,000 per year. The increase in production was attributed to change of variety and fertiliser use. In addition to increased use of improved varieties, increases in productivity were also attributed to the use of fertiliser. Among the 140 surveyed households, the proportion of households using planting fertiliser had increased from 59 per cent to 93 per cent and the proportion of households using top-dressing fertiliser had increased from 15 per cent to 64 per cent.

These results are considerable and they indicate common productivity improvements across all intervention areas. There has not been time to put accurate financial details on the physical results (this is currently subject to further evaluation) but the results presented above give an idea of what the gains might be. For example, in the Keumbu case, using a rough guide price of Ksh 60/kg the average extra value of the maize increase would amount to around Ksh 24,000 or $270. This would equate to over $½ million over 2,000 villages, giving a rate of return that is impressive by any standards. A similar net benefit can be seen in the other case examples.

Enterprise development

The main aim in the RIU Best Bet programme was to explore how best to facilitate the private sector in technology development for the rural poor. In the initial call for proposals this was flagged as an important criterion and successful bids at the 'dragons den' meeting all had very clear statements about how they would set about achieving this. The rationale behind this emphasis for RIU was one of incentives combined with sustainability. All too often technology interventions only last as long as the project aid supporting it; if NGO-led it is often hard for recipients to continue to exploit them sustainably beyond this point. Conversely, at RIU Best Bet project selection, all candidates were asked to demonstrate significant private sector involvement in their applications and to outline an exit strategy showing how impact would continue after intervention ceased.

Table 7.2 shows how successful this has been. All nine projects have had some impact on private sector development and to quite a varied extent. To some degree this has been on core producers like Real IPM and Kenya Biologics who have developed a new type of market for themselves, though both are facing issues of access to geographically isolated markets.[6] In others, as with outgrower activity, it has directly affected income earning in poor farming communities. In the FIPS and SOS cases it has enabled small enterprises to emerge that fulfil an extension role while at the same time expanding markets for established drug and fertiliser suppliers. However, in practically all cases the scale of impact is not yet clear due to the short time scales involved.

Table 7.2 Best Bet enterprise development

Best Bet	*Enterprise Development*
Stopstriga/Gro-plus	Expanded low income market for Real IPM; potential 500,000 for Gro-plus; significant feedback results; engagement of agrovets as distributors; poor results for myco herbicide
BCA Ghana	Expanded markets for Real IPM and KB. Recruitment of private sector distributors in W Africa; potential for export crops (especially cocoa); also groundbreaking document on bio pesticide registration process
Armyworm	Establishment of Biotechnology facility at Arusha with private sector, government and NGO markets; involvement of foreign BCA company setting up local office
Aquashops	Franchising company; fish farms; potential market for fingerling and feed suppliers
NERICA Rice	Outgrowers; expanded seed sector firms; expanded agro dealer business.
Seed Yam	Outgrowers working as seed entrepreneurs
FIPS	New markets for fertiliser and seed firms; small farmer access to private sector technology; VBA businesses created
SOS	New agrovet businesses; improved markets for drugs and insecticides
Shujaaz	Uncertain – geographical extension of concept into other areas of East Africa is the aim

Source: Internal RIU document

For example, in the seed yam case the first planting did not begin until autumn 2011 since access to credit, combined with other difficulties, had hampered out-grower production at early stages although it is expected that the BMGF will now provide further support.[7] Similarly, in the Aquashops case problems with the franchise model along with technical issues have meant production is really only just beginning to take place, while the Armyworm facility was delayed by poor rains in 2011. In the NERICA Rice Best Bet, though considerable progress in the out-grower model has taken place, it has not been possible to see results at branding and marketing levels. In all these cases therefore, it is too early to make judgments but enough promise is present to suggest that further public investment might pay off in the longer term.

In two cases there are signs of productive benefit. In the case of Real IPM there is evidence of considerable benefit using the fertiliser seed priming mix. The claimed benefits of using this two stage treatment (seed priming with a starter fertiliser) are faster emergence of seedlings, stronger and more developed root systems, more vigorous growth, earlier flowering and harvesting, high yields, lower risk of crop failure under drought conditions after emergence and low cost – therefore less risk to investment. Can be used for both farmer saved seed and purchased hybrid seed. Trials have been carried out in Western and Eastern provinces of Kenya under a variety of conditions and the product has shown promising results as highlighted in Table 7.3.

Table 7.3 Comparative improvements of seed priming/fertiliser technology

Product Characteristics	*Without Gro-plus*	*With Gro-plus*	*+ve Change %*
Number of grains	476	500	5
Cob length (cm)	21.7	22.5	4
Cob diameter (cm)	2.4	2.94	23
Grain mass (g)	183.2	223.5	22
Cob mass (g)	43	63.7	48
1000–seed weight (g)	356	423	19

Source: Internal RIU document

Venture capital finance

Finally, it is worth mentioning that at least three RIU projects (including two Best Bets) are now under discussion with financial intermediaries to scale them up and out on a solid and sustainable commercial footing. As we have seen, a major issue in technology development is not so much accessing the underlying scientific knowledge (often well understood) but rather difficulties in actually exploiting this knowledge productively in a given problem situation. Donors are often reluctant to support the 'use' part of technology development and regrettably many scientific bodies are complicit in this, preferring to remain 'at the bench' in their host bodies. But there is now growing interest in public-private partnerships (PPPs) and innovative ways of funding social capital using private sector resources. These include the issuing of social impact bonds where returns depend on demonstrated development results, and the capitalising of those parts of projects where good social and private returns seem likely to occur.

This has now taken place with two of the Best Bets (FIPS and SOS)[8] and one of the Africa country projects. In the case of FIPS a venture capital company is in discussions designed to commercialise production of sweet potato and vegetables. Details are being worked out but are expected to include an agreed share of ownership to FIPS to enable the company to continue its extension activities. In the case of SOS, the consortium is in discussions with private interests to scale up cattle treatment on a long-term commercial basis in Uganda. In this case some risk protection is envisaged through a form of publicly financed insurance. The 'non Best Bet' project concerns the development of a warehouse receipts system in Rwanda designed to improve returns to poor maize farmers. Further details on this project may be found in Gildemacher and Mur (2012). Since discussion on these projects is currently confidential, unfortunately we cannot provide further details at this stage. But reference may be made for example to a relevant website.[9] This provides information on the type of approach that is being taken.

Institutional context

Biosafety regulation

One conclusion that may readily be drawn from some of the Best Bets is the influence of public sector regulation and control which can and does vary across countries. This affected particularly the biotechnology-related projects where in some cases technology development has been held up for significant amounts of time. Thus, in the Stopstriga example, the Kenya Pest Control Products Board (PCPB) restricted registration trials of Stopstriga to the greenhouse in the first instance on bio safety grounds. Unfortunately PCPB protocols for bio safety have been developed only for broad spectrum chemical biocides and these are unnecessarily restrictive for bio control agents that work only with specific pests.[10] In this case Stopstriga was not promoted with farmers as the team did not want to raise expectations if they were prevented from doing so by PCPB. This Best Bet has shown that there is a real issue with getting BCAs registered for use in

Kenya. The registration of Stopstriga has thus been a delay causing a greater focus on Gro-plus seed priming.

At the same time, the delivery system for Stopstriga has had to be revised since it is not offering effective control (much poorer than the results from the field in Ghana where the fusarium isolate originated). Real IPM is now looking for a local isolate that they will source from striga fields in Western Kenya. The value of selecting a Kenyan isolate of Stopstriga (*Fusarium oxysporum*) is that it will not require regulatory permission to do field research in Kenya. The contrast with the BCA project in Ghana is interesting. In this case, the Ghana EPA has clearly been interested in the wider applicability of BCAs in export crops for international markets, particularly cocoa. This, combined with close interaction with private sector organisations, has speeded up registration channels. It seems also to have created a climate supportive to the wider adoption of BCAs in Ghana.

Extension services

As outlined in Chapter 1, there are now serious concerns about how much national extension services in LICs are 'fit for purpose'. As one interviewee put it 'the system is now too cumbersome, under-resourced and top-down to manage what is expected of it'. Recent literature bears this out. For example, Anderson (2008) maintains that 'one can assume based on earlier surveys that typical farmer-to-extension agent ratios are in the range between 1,000 and 2,500 to 1'.[11] When account is taken of the relative lack of experience of extension personnel, time and resource constraints, language difficulties and distrust of them as 'outsiders' on the part of farmers, it would not be surprising if contemporary evidence concluded that agricultural extension workers have so far had little substantial impact on productive practices in rural agriculture.[12]

And indeed it does appear that in many parts of the world, extension agencies are able to do little more than scratch the surface of rural poverty and have to rely on ancillary *ad hoc* inputs from a variety of public and private advisory services. Nevertheless, the RIU programme has shown that extension services can on occasion play a supportive role. In the Armyworm project for example, the Kenya Plant Protection Division were able to solicit extension help in the rolling out of forecasting traps,[13] while in the SOS project DVOs have played a similar role in assisting with cattle spraying. Again, the conclusion must be that existing structures can and should be used where they are able to play a complementary role and where use of private resources would be inefficient. Conversely, where this is not so then recourse to private sector investment will be necessary. The FIPS case appears to be an example where the use of public resources has been able to generate small entrepreneurial extension businesses through the VBA system, while in the Real IPM case recourse to small-scale agrovets (or agro dealers) is seen to be a way forward to promote its new fertiliser technology.

Tertiary education

When RIU was conceived, there was a particular emphasis on working with the holders of the tacit knowledge which underpinned research in the RNRRS programmes; this meant that significant links were activated with UK universities from the outset. And we have seen at a programme level how links with European universities have been further developed (see Table 7.4).

Table 7.4 European universities

University	*Best Bet*	*Activity*
University of Edinburgh, UK	SOS Uganda	Participant in the work programme
University of Edinburgh, UK	SOS Nigeria	Participant in the work programme
University of Edinburgh, UK	RIU Programme	University of Edinburgh has set up a subsidiary company, *Research Into Results*, to manage the overall programme and exploit learning to date
University of Exeter, UK	FIPS Africa	Dissemination of reduced tillage techniques for improvement of soil health and water harvesting
Open University, UK	Two staff: (i) Senior Adviser to the programme and (ii) leader of the CRT	Technical backstopping and training Research Fellows – in some cases to PhD level
University of Greenwich, UK	Real IPM Company – Stopstriga and BCA (Ghana)	Technical backstopping – biological control agents
University of Greenwich, UK	Armyworm	Technical backstopping – biological control agents and forecasting technology
University of Lancaster, UK	Armyworm	Technical backstopping – biological control agents and forecasting technology
University of Hohenheim, Germany	Real IPM Company – Stopstriga and Bio-pesticide registration (Ghana)	Technical backstopping – biological control agents
University of Bangor, UK	Real IPM Company – Gro-plus	Technical backstopping – biological control agents
University of Stirling, UK	Aqua Shops	Technical backstopping – seed priming
University of Surrey, UK	Clean yam planting material	Technical backstopping

Source: Internal RIU document

However, as the work has progressed it appears that projects have increasingly looked to African universities for practical support. Table 7.5 shows this from the standpoint of the whole RIU Africa programme. This was in part a reflection of change in emphasis over time. Initially, DFID-funded RNRRS research was prioritised over other potential solutions. But gradually a more pragmatic approach was adopted which mixed the RNRRS research with research generated elsewhere.

Table 7.5 RIU support from African universities

University	*RIU Africa Country Programme*	*Activity*
University of Zambia	Zambia	Support on holding a symposium on agriculture
Mzuzu University, Malawi	Malawi	Student exposure to RIU programme and practical inputs to course
University of Malawi	Malawi (Fish innovation platform)	Research on improved fish strains and on fish feed formulation within context of small scale fish marketing studies
University of Malawi	Malawi (Livestock innovation platform)	Provided the project champion to aid capacity building of other livestock farmers
University of Malawi	Malawi (Legume innovation platform)	Variety of legume released
Bayero University, Kano, Nigeria	Nigeria (Legume innovation platform)	Training for agricultural extension workers in storage of cowpea
Ahmedu Bello University, Nigeria	Nigeria (Crop innovation platforms)	Improved varieties of high-yielding dual purpose seed for cowpeas; soybean seeds and cassava cuttings along with training for agricultural extension workers
Makerere University, Uganda	Rwanda	MSc Thesis on Innovation Platform development in Rwanda
University of Unatek, Rwanda	Rwanda	Internship on cassava innovation platform

Table 7.5 continued

University	*RIU Africa Country Programme*	*Activity*
Umutara Polytechnic, Rwanda	Rwanda	Thesis on warehouse receipt systems
Njala University, Bo Sierra Leone (Milton Margai College of Education and Technology)	Sierra Leone (Poultry innovation platform)	Cataloguing research local research results
School of Hygiene, Sierra Leone	Sierra Leone Solar Fruit Drying innovation platform	Production of manual and training materials on solar drying
Fourah Bay College, Freetown, Sierra Leone	Sierra Leone	Media survey
University of Dar es salaam, School of Journalism, Marketing and Communication	Tanzania	Development of a module designed to improve the quality of writing on science and development amongst journalists

Source: Internal RIU document

RIU is also having an impact on university teaching. We can see in Table 7.6 (drawn from Table 7.5) a number of Best Bet examples where undergraduate and post-graduate students have been exposed to opportunities to gain practical experience in the field. This means that, through engagement with RIU, students leaving African universities are more work-ready and have a better understanding of the expectations of employers in practical business and development contexts.

The clearest example of this may be seen from the SOS project where an arrangement has been made with Makerere University which to our knowledge is fairly unique in Africa. The university's new *Institute for Strategic Animal Resource Services (AFRISA)* has been institutionalised as a company limited by guarantee. Facilities at the veterinary faculty have been rehabilitated, and equipped to provide its secretariat. Later, as the initiative gains momentum, AFRISA will be expanded and upgraded into a prime strategic centre for development of graduate survival skills and community service for the animal industry. The planned skills training and service opportunities provided by the institution will respond to the strategic and practical skills requirements of the animal and public health industry in Uganda, and will be harmonised to blend with relevant University and partner programmes. With the FVM AFRISA Centre as the fulcrum, the university plans to expand and integrate to other disciplines and units that find benefit in the initiative both nationally and internationally. The actual amount and cost of work will be determined by technical needs assessments.

In this way AFRISA is envisaged as an enabling facility to promote student placement and entrepreneurship activity starting with the SOS project but growing eventually to impinge on all aspects of Makerere education. It will be an independently financed institute housed within the university but with the capacity to create *ad hoc* partnerships as needed. In the SOS case such a unit has established a partnership with relevant private

Table 7.6 LIC university capacity building

Programme	*University*	*Activity*	*Capacity building*
Aqua Shops	Moi University, Kenya	Fish farming research	Development of a syllabus on fish farming business
BCA	University of Ghana – Crop Science	Bio-pesticide registration in Ghana	Efficacy trials on bio-pesticides
Seed Yam	University of Ibadan, Nigeria	Nematology Research on pesticides residue in yams	Training in seed yam production
Real IMP/GroPlus	Kenyatta University, Kenya	Efficacy of phosphate in priming	Managed with MSC students
FIPS	Kenyatta University, Kenya	Recruited students to do work in vacations; impact assessment work	Students gain practical placement experience
FIPS	Moi University, Kenya	Recruited students to do work in vacations; impact assessment work	Students gain practical placement experience
SOS	Makerere University, Uganda	Student field work activity	Development of AFRISA – the university-delivered training programme working along agricultural value chains

Source: Internal RIU document

sector and public sector bodies that have an interest in veterinary employment of graduates. It has also helped to establish a mechanism to promote the creation of relevant private sector business opportunities and related facilities. Such opportunities will benefit from inputs from external technology and finance but will ultimately aim to be self-sustaining and independent of external support except where public interest is required. In time, the lessons learned will be used to move into other areas of potential graduate employment. In summary therefore, AFRISA is a capacity-building measure focusing on a 'one-health' research/extension model that will be instrumental in delivering:

1. a student-based extension service system
2. a technology delivery system
3. a pragmatic incubator for graduate-based business and job creation
4. a vehicle for curriculum development and implementation
5. a vehicle for community development.

Networks and linkages

Finally it is worth commenting on the ways in which RIU has built up networks of actors associated with its interventions. This actually started at the 'dragons den' inception meeting where a conscious effort was made to treat all short-listed applicants as a homogenous group. In the day preceding the interviews, the applicants were given instruction on how to pitch their initiative in such a way as to impress the Dragons with the sustainable viability of projects. They were also told that the procedure was not internally competitive since success would depend entirely on each project's perceived merits. As a result, each Best Bet came to know in detail about all the others and this initial interaction has led to subsequent co-operation among projects. There are many examples of this. One is the interaction between FIPs and Real IPM where the FIPS extension model has been used in assist the scale out of GroPlus. Another is the series of 'stories' Shujaaz portrayed in its comic series.

While the funded consortia acted as 'networks' with each agent providing its own specialist input, technology development in all cases involved a lead 'champion' who took responsibility for driving projects. In many cases project development then began to make links with other programmes involved in related activities as with Armyworm and DLCO in Ethiopia, and Aquashops with Indian counterparts. There are two aspects of this that are relevant from a policy standpoint. The first is that far from being straightforward, putting research into use turns out to be a process of great complexity. In the Armyworm, BCA and Real IPM/Stopstriga cases for example, problems of government regulation have clearly held up progress. This has placed a premium on the effective management needed to broker successful technology development.

Secondly, all the Best Bets acted as nodes of continuous learning and project adaptation to evolving circumstance. It follows therefore that projects of this type cannot really be handled as 'stand alone' activities with predetermined outputs. Instead, they are really heuristic interventions that inevitably need adaptation as they proceed. In cases like the Stopstriga experiment, issues of regulation combined with poor initial outcomes forced the project to focus more on seed priming. In the Aquashops case, technical and structural problems delayed progress. Of course issues of this kind are well understood in industrial development where business firms routinely expect new ventures to undergo teething activity (as we stressed in Chapter 2). Perhaps, however, technology development aid projects in agriculture need to take a similar perspective.

Conclusions and policy recommendations

The opening chapters of this monograph began by setting the DFID RIU experiment in the wider context of agricultural development in low income countries. Despite the great advances made by the Green Revolution the present status of such countries is still dire and this is despite over 50 years of funding on the part of international agricultural science. For DFID, the immediate issue in 2005 was concern that significant sums of public money spent in one of their flagship programmes (RNRRS) were not apparently achieving much in development impact terms. Thus it proved hard to identify examples where the target communities, low income farmers in LICs, were benefitting. In our view, the issue that DFID had identified is really an example of a wider structural problem – that of knowledge market inefficiency – a market failure. By this we mean that 'supply of' and 'demand for' scientific information does not normally correspond. There is a mismatch and so knowledge markets remain inefficient and (often considerable) waste can occur. In turn this means also that science funding, despite decades of attempted reform, continue to be driven by criteria that place a low premium on how this knowledge contributes to social welfare.

The DFID response was a very practical one – to facilitate practical application of its (previously funded) research and to achieve better understanding of how this facilitation might be improved. The impetus stemmed from the apparent failure of technologies derived from previous RNRRS research, to be actually adopted in practice. This was despite successful project completion in a purely scientific sense (i.e. in published papers and associated documents). The Best Bets outlined in this monograph, however, were only put in place in 2009 following a change in RIU management. They all specified private sector involvement since it was hoped this would promote longer term sustainability once public aid support has ceased; the rationale was that donor aid is necessary to cover the risks associated with pre-competitive social costs of technology development but often acts as a disincentive to longer-term entrepreneurship. A major working hypothesis therefore was that technology development needs a further impetus from private sector players since research outputs have usually remained 'on the shelf' in the absence of further support. Since at the start of the programme it proved hard to identify examples of 'knowledge use', RIU explored different mechanisms to facilitate greater use and one of these was the Best Bet sub-programme. Although many of the projects are still on-going there are, we believe, a number of preliminary conclusions that can be drawn:[14]

1. The first point to make is that despite an early failure to identify 'low hanging fruit,' a total of 61 RNRRS projects were 'put into use' by the RIU Best Bet process.
2. The selection process, combined with flexible technical and financial management, enabled the Best Bets Programme to develop the most appropriate pathways to its objectives (in full consultation with RIU management) but without being constrained by rigid management tools. This indicates perhaps the advisability of a lighter touch to be taken in relevant technology development aid.
3. In many cases the original scientists and their organisations continued to play an important role as mentors and advisers in subsequent technology development. This stemmed from their tacit knowledge of the problem area and many years experience in the field.
4. In all cases, however, the mobilisation of other linked knowledge sources proved necessary. Often these derived from NGO bodies but included government departments, other international science bodies, local scientific institutions, as well of course as the private sector itself. Indeed, the projects worked best as consortia in which the different stakeholders operated as an holistic innovation system with each player contributing its own unique expertise.
5. In all cases RIU project funding played a necessary role in covering pre-investment costs associated with risk and related factors such as exemplified in the social costs of Armyworm forecasting. In our view it is likely that this type of pre-competitive support will continue to be an area for necessary technology development aid.
6. The actual business of technology development was complex in all cases; it involved applications engineering, negotiations with government regulatory bodies, accessing products through imports (in the absence of local production capacity) and dealing with the many problems that always plague new innovative ventures.
7. In all cases private sector interests have played a key role, both as 'product champions' and as a core activity ensuring economic continuity. Clearly, economic incentives have an important role in ensuring longer term sustainability and in some cases markets for established firms have expanded as new outlets have been created. In addition, there have been a range of new small businesses created as a

result of RIU interventions and there are indications that new forms of financial support will be forthcoming.

8. The evidence has confirmed the generally held view (outlined in Chapter 1) that formal national government led extension systems need to be reformed and it is our view that the private sector should now play a much greater role in this respect. In this way, it will help to improve the pace and impact of technology development for the rural poor.
9. In some of the projects examined, national regulations and their application have proved a significant constraint though there is also some evidence of reluctance of farmers to use bio control methods.[15] This seems mainly due to a lack of suitable guidelines for bio control agent use. Countries still use guidelines designed for the use of chemical biocides and have difficulty making appropriate judgements. The issue, however, varies across countries with, in the BCA case, Ghanaian regulators proving much faster to adapt than the Kenyan equivalent.
10. In many cases, RIU Best Bet projects have helped to mobilise national capacities, particularly in universities. This is important in the light of frequent criticisms of TE sector viability in Africa and the need to encourage local innovation. The RIU experience appears to show how higher education might play a more substantial role in economic development.

On the basis of these cases, it is clear that 'putting research into use' is by no means something that occurs spontaneously. But this does not mean that past DFID research has been a wasted effort. Far from it; we have shown (albeit on a small scale initially) that given the right networks and environment, much valuable research can be put to use in the developing world. In the context of LIC agriculture, it is a complex process that needs to operate and be managed as a necessary development activity. And since public resources are involved, this means developmental aid. There is therefore a continued need for support to related science and technology activity. However, on the basis of the Best Bet experience we do not believe that bilateral agencies such as DFID should continue traditional funding patterns at current levels by way of handing out grants and standing back. Simply funding universities and other research institutions to conduct yet more disinterested research is by itself insufficient. Rather what is needed is investment on a sustainable basis that ensures this knowledge is actually put into developmental use. In the RIU case, the Best Bets Programme applied a hands-on operating approach. Rather like a 'private equity' firm seeking value for its investors the programme sought value for DFID investments in research. When this environment was established, successful outcomes took place.

What this implies for the wider science policy agenda is a subject for creative analysis. Certainly the conventional *mode 1* approach outlined in

Chapter 2 is insufficient. Following this approach, research-based knowledge would continue to remain 'on shelves' and contribute only marginally to development. Part of the problem is clearly a lack of the business skills needed to establish innovative ventures. But public support to alleviate commercial risks still requires complementary input from the scientific community. We suggest that current UK moves to integrate institutionally research council activities with overseas aid is probably the right way to go. But it will need a pattern of appropriate incentives that encourages scientists not only to undertake applications engineering research but also to link more closely to other bodies involved in practical development, including especially finance and private enterprise. If this is too much for scientific bodies to accept at once, DFID could continue to fund cognate RIU activity on an experimental basis until such time as this more systemic approach becomes more widely accepted.[16] It is our firm belief that a policy shift along these lines will create better 'value for money'. In other words, it will improve the efficiency of the knowledge market in the context of low income country development.

Notes

1 DFID (2006). Strategy for Research on Sustainable Agriculture (SRSA, 2006 to 2016). RIU design considerations are further discussed in www.oecd.org/dataoecd/7/7/35242503.pdf.

2 The idea of the database came about because RIU management were unable to decide on a smaller target number (30) of Best Bets. Further consultation with the original RNRRS programme managers led to the 280 figure though some managers wanted a database of all 1,600.

3 And some of the outputs from these projects were often informed by earlier projects which had increased the knowledge base.

4 It is worth mentioning that some of this funding was from the bilateral vote (not research) which was very much about capacity building via seconded technical co-operation officers (TCOs).

5 What follows is a summary drawn from internal project documents. A more detailed set of evaluations may be accessed through Gildemacher and Mur (2012).

6 For example, access to isolated agrovet shops is proving a problem to the Real IPM and they are now considering operating through larger companies. There are also issues about differential national market conditions.

7 This has now taken place. See http://allafrica.com/stories/201204021448.html. Incidentally, it also shows that RIU has had success in leveraging additional funds.

8 A third project, Armyworm, is still being considered.

9 See www.socialfinance.org.uk/work/sibs.

10 In fact biological control protocols are laid out in this document: www.dfid.gov.uk/r4d/PDF/Outputs/. ../R7299Registration.pdf. At the time, this RNRRS funded initiative was lauded as groundbreaking and opened Kenya up for the registration of BCAs. However the RIU experience is that since that time the authorities have been overly strict in the interpretation of the protocols.

11 See Anderson (2008: 11). Anderson uses earlier surveys as sources for his estimate.

12 See also, for example, Feder *et al.* (2010) and Rohrbach (2010). Anderson (2008) also provides a useful outline of the demise of the World Bank T&V system after c.1995; see p12.
13 And it is worth noting that one trap can serve communities of several thousand.
14 For a more detailed analysis of the Programme as a whole (which includes two of the Best Bets – FIPS and Armyworm control) see Gildemacher and Mur (2012).
15 The other problem is that bio control agents take a little longer than chemical ones to function effectively and farmers are reluctant to wait the extra few days.
16 Indeed scientists are often keen to get involved. Gildemacher *et al.* (2012) explore the case of a seed selection programme in east Africa where the scientists developed an approach to improve the quality of seed potatoes by ware potato growers and were encouraged to do this by their institutions. This approach is complementary to specialised seed production systems and is now widely promoted in sub-Saharan Africa. See p. 1.

Appendix 1: RIU discussion paper series

This series may be accessed at: http://researchintouse.com/learning/learning40discussionpapers.html. See also http://www.researchintouse.com/resources/riu10uk-appg-agfood-brief.pdf.

Discussion paper 01
Research into Use: Investigating the relationship between agricultural research and innovation
Authors: Andy Hall, Jeroen Dijkman and Rasheed Sulaiman V
July 2010

Discussion paper 02
Bottom-up, bottom-line: Development-relevant enterprises in East Africa and their significance for agricultural innovation
Authors: Andy Hall, Norman Clark and Andy Frost
July 2010

Discussion paper 03
Innovation systems, economic systems, complexity and development policy
Author: Norman Clark
September 2010

Discussion paper 04
Putting research into use: A market failure approach
Author: Norman Clark and Ian Maudlin
September 2010

Discussion paper 05
It may take a little while ... Insights on agricultural research for innovation and development in Nigeria
Author: Utiang P Ugbe
October 2010

Discussion paper 06
Gender and agricultural innovation: revisiting the debate through an innovations systems perspective
Author: Ann Kingiri
October 2010
Related paper: Rethinking gender in agriculture innovation from an innovation system's perspective
Publisher: African Centre for Technology Studies Nairobi, Kenya as a policy brief
Authors: Ann Kingiri, Judi Wakhungu and Andy Hall
December 2011

Discussion paper 07
New organizational and institutional vehicles for managing innovation in South Asia: Opportunities for using research for technical change and social gain
Authors: T S Vamsidhar Reddy, Andy Hall and Rasheed Sulaiman V
October 2010

Discussion paper 08
The innovation trajectory of sleeping sickness control in Uganda: Research knowledge in its context
Author: John Morton
October 2010

Discussion paper 09
Africa Matters: emerging lessons from the RIU Country Programmes
Author: Jeroen Dijkman
November 2010

Discussion paper 10
What does innovation smell like? A conceptual framework for analyzing and evaluating DFID-RIU experiments in brokering agricultural innovation and development
Author: Utiang P Ugbe
November 2010

Discussion paper 11
Studying rural innovation management: A framework and early findings from RIU in South Asia
Authors: Rasheed Sulaiman V, Andy Hall, T S Vamsidhar Reddy and Kumuda Dorai
January 2011

Discussion paper 12
Organized retailing of fresh fruit and vegetables: Opportunities for putting research into use?
Authors: Rasheed Sulaiman V, N J Kalaivani, Jatinder Handoo, T S Vamsidhar Reddy, Kumuda Dorai and Andy Hall
May 2011

Discussion paper 13
Beyond knowledge brokerage: An exploratory study in innovation intermediaries in an evolving smallholder agricultural system in Kenya
Authors: Catherine W Kilelu, Laurens Klerkx, Cees Leeuwis and Andy Hall
May 2011

Discussion paper 14
The when and the where of research into agricultural innovation trajectories: Evidence and implications from RIU's South Asia projects
Authors: T S Vamsidhar Reddy, Andy Hall and Rasheed Sulaiman V
June 2011
Related paper: Locating research in agricultural innovation trajectories: Evidence and implications from empirical cases from South Asia
Publisher: Oxford University Press – the Journal of Science and Public Policy
Authors: T S Vamsidhar Reddy, Andy Hall and Rasheed Sulaiman V
Publication date: to be announced

Discussion paper 15
Dynamics of bioscience regulation and opportunities for bioscience innovation in Africa: Exploring regulatory policy and brokering
Authors: Ann Kingiri and Andy Hall
June 2011

Discussion paper 16
Necessary but not sufficient: Information and communication technology and its role in putting research into use
Authors: Rasheed Sulaiman V, Andy Hall, N J Kalaivani, Kumuda Dorai and T S Vamsidhar Reddy
June 2011
This paper is scheduled for publication in the *Journal of Agricultural Education and Extension*
Date of publication: unknown

Discussion paper 17
Functions and forms of brokerage in the Malawi fisheries platform
Author: Elias Madzudzo
July 2011

Discussion paper 18
Embedding research into use ideas in the policy space: The case of RIU Nigeria and Sierra Leone
Author: Utiang P Ugbe
July 2011

Discussion paper 19
Brokering in practice: The experience of the RIU Malawi Country Programme
Author: Maija Hirvonen
July 2011

Discussion paper 20
Research Into Use: An institutional history of the RIU Nigeria country programme
Author: Maija Hirvonen
July 2011

Discussion paper 21
NERICA seed versus local landraces: Another battle of the paradigms?
Author: Maija Hirvonen
August 2011

Discussion paper 22
Emerging development-relevant enterprises in Kenya: Do they exist, what do they look like and what is their role in poverty alleviation?
Author: Andrew O Adwera
August 2011

Discussion paper 23
Exploring mechanisms for putting research into use: Evidence from RIU's value chain-oriented projects in South Asia
Authors: TS Vamsidhar Reddy, Rasheed Sulaiman and Andy Hall
August 2011

Discussion paper 24
Beyond Biosafety Regulation: Implications for putting biotechnology research into use in a developing country context
Author: Ann Kingiri
September 2011

Discussion paper 25
Missing the target: Lesson from enabling innovation in South Asia
Authors: TS Vamsidhar Reddy, Rasheed Sulaiman V and Andy Hall
September 2011

Discussion paper 26
Putting Research into Use: A market failure approach
Authors: Norman Clark, Andy Frost, Ian Maudlin, Paul Seward, Henry Wainwright and Andrew Ward
September 2011
This paper has now been published as Clark, N., Frost, A., Maudlin, I., Seward, P., Wainwright, H. and Ward, A. (2011), 'Putting Research into Use: A Market Failure Approach', *International Journal of Technology Management & Sustainable Development,* 10(3): 185–200.

Discussion paper 27
Putting Research into Use: Lessons from contested visions of innovations
Author: Andy Hall
April 2012

Appendix 2: RNRRS projects

The following were the main RNRRS projects that were planned to be put into use at the start of the Best Bets. They are written as a summary of their primary objectives and content and are listed under the main RIU projects they related to at the time of inception of these projects. However, where they were used in other RIU projects, this is also noted.

Armyworm

R6746 Entomopathogenic viruses for control of African Armyworm (Spodoptera Exempta) in Tanzania

The purpose of this project was to develop and evaluate the use of a natural biological agent, an insect virus, *Spodoptera exempta nucleopolyhedro* virus (SpexNPV), as a safe alternative to chemical insecticides for control of the African Armyworm (*S. exempta)* in Tanzania. The project achieved its key output of producing and field testing SpexNPV as a control tool for African Armyworm. However some important questions remained to be addressed before it is ready to be promoted to farmers on a large scale. It needs to be tested on a full range of cereal crops in farmer participatory trials to ensure it is compatible with existing farmer practices. Local production techniques need to be developed, evaluated and costed. There also needs to be socio-economic analysis to determine if this new technology can be successfully integrated into the existing national and regional Armyworm control.

R6762 Decision tools to aid Armyworm surveillance and outbreak prediction (also Shujaaz)

Demand for the project was identified by national and international agencies involved in Armyworm control. The purpose of the project was to develop population and behaviour models of economically important Armyworm outbreaks and to promote improved pest management strategies. This was achieved through the following activities: (1.) The development of a model to describe the dispersal, displacement and re-concentration of Armyworm moths; (2.) The development of a mechanistic, rule-based, model of Armyworm population dynamics and movement; (3.) An analysis of the behaviour of the model (2. above) for a representative range of initial conditions and weather scenarios, and (4.) The dissemination of findings, and training in use of the forecasting aids. These activities have contributed to DFID's development goals by reducing the impact of migrant pests on crop production, as more accurate outbreak prediction improves the efficiency of control teams.

R7954 (2001–2004) Novel technologies for control of African Armyworm on smallholder cereals in East Africa

This project, in collaboration with the Tanzanian Ministry of Agriculture, sought alternative novel controls based upon the use of a natural pathogen of Armyworm, the *Spodoptera exempta nucleopolyhedrovirus* (SpexNPV) and a local botanical neem. The project demonstrated that SpexNPV can be used to control Armyworm outbreaks as effectively as chemical pesticides. This has led to important new insights into host/virus relations in migratory insects and may be of major importance in understanding Armyworm population dynamics. SpexNPV isolates have been characterised, and study of their genetics and biology is continuing with UK research council funding. The Tanzanian Ministry of Agriculture has directed its Armyworm control agency to develop a strategy to adopt SpexNPV in place of chemical control. It is also implementing and evaluating the use of neem in a community Armyworm forecasting and control project co funded by the USAID and the Government of Tanzania.

R7966 Identifying the factors causing outbreaks of Armyworm as part of improved monitoring and forecasting systems (also Shujaaz)

The project aimed to provide new forecasting tools for effective Armyworm control employing limited resources. The national system requires, *inter alia*, real-time, spatially accurate rainstorm information derived from Meteosat images. A socioeconomic survey assessed whether farmers would be interested in carrying out their own Armyworm forecasting (using a rain gauge and pheromone trap located in the village), and in what way they are willing to respond to forecasts, either local or national. Community-based pilot schemes were set up in five villages in an Armyworm 'hot-spot' village. The design of the forecasting was shared by the communities and during the first year of the pilot study, traps were successfully run and forecasts made in all of the participating villages. A forecasting pack, in both Kiswahili and English, was produced to accompany training in community forecasting. This includes information about Armyworms, instructions on how to operate the pheromone trap, the rain gauge, and how to record and interpret the data to make the forecast.

R8407 Improved Armyworm monitoring and forecasting systems (also Shujaaz)

Community-based forecasting (CBF) represents a significant paradigm shift in migrant pest forecasting. The prevailing view was that migrant pests, by their nature, are international and therefore their forecasting can only be tackled by centralised organisations. However, forecasting at a village level can be both feasible and complementary to the national service. The Armyworm problem is important, widespread and farmers want to do something about it. The CBF approach used village level technology and farmers have seen that it works. This is evidenced by the fact that pilot villages continued to run CBF. In addition, local initiatives took place as a result; e.g. a sprayer rental scheme and a contract sprayer group. As a sign of institutional investment, in some cases both village and district authorities allocated funding to allow CBF to continue. Pilots were then carried out in Kenya and Tanzania as well as an initial pilot in Ethiopia. The next challenge is to devise effective ways to present greater numbers of villages the opportunity to develop CBF.

R8408 (2005) Novel technologies for control of African Armyworm on smallholder cereals in East Africa

Existing controls based upon chemical insecticides are rarely able to meet more than 30 per cent of the needs of poor farmers and subsistence growers due to high cost and limited availability. The SpexNPV application technology and project findings were demonstrated to national Armyworm control scientists from Kenya and Malawi at regional workshops. Regional Armyworm control organisations DLCO and FAO were also briefed. The director of DLCO agreed to champion SpexNPV technology and seek funding to scale out the technology to other Armyworm affected countries in Eastern and Southern Africa. The project team were actively engaging with national and regional Armyworm control agencies and donors to achieve scale up and scale out of SpexNPV in SSA so that the benefits of improved Armyworm can be realised by the large numbers of poor farmers and households whose income and food security this pest threatens.

Stopstriga/seed priming

The myco herbicide used to control striga had already been developed by the University of Hohenheim though the African Enterprise Challenge Fund (partly funded by DFID) providing part grant/part loan assistance for the registration of ‘Stopstriga’. Also IDRC funded basic research into the myco herbicide. The following RNRRS projects provide the scientific knowledge upon which the seed priming element of the 'Stopstriga' package is based.

R6395 The development and testing of seed-priming to improve stand establishment, early growth and yield in semi-arid Zimbabwe and India

Based on earlier DFID-funded work in southern Africa, the concept of 'on-farm' seed priming was developed and tested using a combination of in-vitro, on-station and on farm participatory trials. Interaction between researchers and farmers was a critical factor in the success of the project. Work was initiated in Zimbabwe and India, in marginal areas where seed soaking was a known practice but was not pursued except on a conditional basis, e.g. when sowing was late or otherwise forced to take place under sub-optimal conditions.

R7440 (99–03) Physiological basis for the effects of on-farm seed priming in tropical crops: interactions with seedbed physical conditions

The overall aim of the project was to test the crop physiology underpinning 'on-farm seed priming'. It was established that previously reported benefits of priming in the field could be demonstrated in controlled environments for maize, rice and wheat. In experiments in the laboratory, controlled environments and in the field, there was no evidence that priming inherently changed the way that plants grow. Instead, the benefits depend on the extent to which priming can advance germination and emergence. This in turn depends on the soil physical conditions at sowing. These outputs provide assurance to the development community that priming is a scientifically sound practice that can improve yields through increased crop stand, more rapid seedling emergence and increased uniformity of the stand.

R7438 Participatory promotion of on-farm seed priming

There is ample evidence that poor crop establishment is a widespread constraint of crop production in developing countries, particularly in the marginal environments farmed by poor people. Patchy plant stands are common, and yields are often reduced simply because there are not enough plants in the field. In addition, plants that do eventually emerge often grow slowly, and are highly susceptible to stresses such as drought, pests and diseases. Farmers can choose to re-sow, although this entails severe yield penalties and increased labour and financial costs, and there is evidence from India that borrowing to pay for replacement seed can initiate or add to a spiral of indebtedness. Clearly, anything that can be done to increase the proportion of seeds that emerge will have a large impact on farmers' livelihoods. Low, unstable yields are a major contributor to the fragile lives of poor farmers in marginal areas. Low-cost, low-risk interventions that increase and stabilise yields will have a large impact on the livelihoods of such farmers. This project promoted a workshop designed to promote registration of bio control agents legislation coming out of its horticultural projects in Kenya. The result of the workshop was a legal notice which was the basis for the current bio control legislation.

R7189 Cultivar competitiveness and interactions with on-farm seed priming for integrated weed management

Studies were undertaken by a number of collaborating institutions to investigate aspects of the role of genotype and interactions with pre-planting seed hydration (seed priming) on the weed competitiveness/tolerance of maize and rice. Laboratory experiments at the University of Wales confirmed that priming hastens time to germination of both crops. The practice is thought by farmers to improve crop emergence. Some consider that young primed maize plants grow faster and are more competitive with weeds. Priming is a low cost practice that provides an opportunity for marginal increases in maize yield. In a series of field and greenhouse experiments undertaken at the West Africa Rice Development Association, the effects of seed priming on germination, emergence, early plant growth, grain yield and competitive ability of upland rice were investigated. The studies concluded that although seed priming has a positive impact on early germination, subsequent effects on final emergence and yield are variable.

Bio control agents in Ghana

R7249 Development of mycoinsecticides and pheromones for cocoa mirids

In this project a local isolate of an entomopathogenic fungus active against cocoa mirids was isolated and characterised, produced on pilot scale and formulated so that it was available for field evaluation. Pheromone components for both species of mirids were isolated, identified and synthesised. Blends were shown to attract one of the mirid species in the field and effective traps and dispensers developed for further evaluation. Progress has been slower than expected. Entomopathogens were much more difficult to find than anticipated from work on related species in other countries. The Ghana mirid species still cannot be reared in large numbers and they are very difficult to maintain and transfer between laboratories without heavy mortality. Prior to the project, *D. theobroma* was thought to be the dominant species, but now it is difficult to find compared with *S. singularis*. The pheromone work was hampered by the very small amounts of pheromone produced and the novel structures involved. Nevertheless, the project outputs have provided new technologies that can now be evaluated for use in control of cocoa mirids in Ghana and neighbouring countries. Mirids were unanimously rated as one of the two main pest and disease constraints on cocoa production throughout West Africa at a subsequent workshop 'West African Regional Cocoa IPM'. These bio rational approaches will help minimise use of conventional pesticides and should be more appropriate for use in developing countries and more sustainable. Cost-effective improvements in productivity of cocoa should benefit the livelihoods of over six million people in Ghana alone, who depend directly or indirectly on cocoa.

R7441 Development of pheromone trapping – monitoring and control of the legume podborer by farmers in West Africa

The legume podborer, *Maruca vitrata*, is a major pan-tropical pest of legume crops and particularly of cowpea in West Africa. The objectives of this project were to complete optimisation of pheromone traps and lures that had begun under a previous project; it also aimed to integrate their use with other novel IPM technologies to provide improved methods for control by small-holder cowpea farmers in West Africa. The project was expected to provide a better understanding of the population dynamics, ecology and behaviour of *M. vitrata*, based on long-term monitoring with pheromone traps. Progress has been made in developing pheromone traps to assist in the control of *M. vitrata* by acting as predictors of infestations, enabling the timing of control measures to be optimised. There are strong indications that they will prove useful to farmers in Benin and Ghana. Trials of the trap-threshold concept, in combination with botanical pesticides, have demonstrated the potential of such an approach.

R7449 Developing an improved sustainable integrated pest management system for brassica production in Kenya

This project was designed to develop improved non-chemical insecticide approaches for the management of Kenyan vegetables. The project focused on control of the diamondback moth (DBM *Plutella xylostella*) identified as the priority pest problem for vegetable farmers in Kenya. The objective was to develop and evaluate two alternatives to chemical insecticide, a natural pesticide based upon an insect virus and a pheromone mating disruption technique. After extensive development trials the pheromone mating disruption was found not to be a feasible option for small vegetable farmers in the Kenya. However, the development of an endemic *P. xylostella (PlxyGV)* granulovirus, a virus first isolated in 1997 as a biological pesticide, made very significant progress. A plant to produce *PlxyGV* is now being set up in Kenya with a commercial partner and will be the first viral insecticide to go into commercial production in sub-Saharan Africa. The product will be promoted to provide a safe, effective alternative to the increasingly expensive and unreliable chemical pesticides.

R7960 Public-private partnerships to develop viruses as bio-insecticides for Ghana and Benin

This project sought to enhance incomes of smallholder farmers and to increase food safety through the promotion of viral bio pesticides as alternatives to broad spectrum synthetic insecticides for control of lepidopteran pests of vegetables through public-private sector partnerships in Ghana and Benin. The project focused on the diamondback moth, *Plutella xylostella* as a pest of cabbages. Opportunities for the promotion of bio pesticides stem from rising concern over the safety of chemical pesticides and include the national vegetable IPM strategy in Ghana, new export markets and implementation of MRLs (maximum residue limits) on agricultural produce destined for export, resistance management, organic agriculture, public health campaigns and the Stockholm Convention to replace persistent organic pollutants. Stakeholders agree that commercialisation is the most appropriate implementation route, but see a need to establish specific registration guidelines and local production facilities, to undertake a comprehensive market study and a training programme. Wide-scale adoption of *Plutella xylostella* granulovirus in Ghana and Benin is unlikely without economic incentives or donor support unless these constraints can be overcome.

R8300 Implementing pheromone traps and other new technologies for control of cowpea insect pests in West Africa

This project replicated on-station trials for control of cowpea insect pests in Benin and Ghana building upon earlier results. They showed that a variety of botanical pesticides, with or without pheromone traps for the legume podborer, *Maruca vitrata*, gave yields and infestations which were generally intermediate in effectiveness between conventional pesticides and untreated controls. Neem seed oil was the best botanical medium for controlling flower thrips and *M. vitrata*. Farmer field school (FFS) trials largely confirmed these findings and recommendations were developed for the optimal use of traps. Feedback from farmers and researchers indicated that the use of traps, with a flexible approach to choice of control agent, enabled better integration with existing farmer practice and better understanding and interest among farmers. Studies of the social and economic feasibility of technologies showed that a substantial proportion of farmers would be willing to pay the estimated cost of traps and lures. Progress to develop pilot-scale systems of manufacture and distribution of pheromone traps and lures, and of botanical insecticides, was mixed.

R8313 Cocoa IPM in West Africa

This was a generic project designed to investigate possibilities for integrated pest management in cocoa production and storage (see also R8430 below).

R8430 Development of bio pesticide registration and risk assessment guidelines for Ghana

This project developed bio pesticide registration guidelines and data requirements for promulgation into national legislation. The project also added value to R8300 (Pheromones of *M. vitrata* for cowpea in West Africa) and R8313 (Cocoa IPM in West Africa). In a collaborative effort between the Ghanaian Environmental Protection Agency (EPA), the consultancy organisation JSC International, and the Natural Resources Institute (UK) the project successfully prepared guidelines for applicants wishing to register microbial and biochemical bio pesticides in Ghana. In addition to drafting guidelines, training was provided in bio pesticide risk assessment and data evaluation to members of the EPA's Chemical Control Management Centre (CCMC), the Pesticide Technical Committee and its sub-committees. Training gave staff an understanding of the specific, yet often qualitative issues involved in registering bio pesticides, compared with traditional synthetic pesticides. Through its research programmes, DFID has demonstrated a commitment to the development of environmentally friendly crop protection. Preparation of these guidelines addresses one of the recognised constraints to development of bio pesticides highlighted in the recent Crop Protection Programme-funded project R7960. While the project specifically targeted the Ghanaian EPA, observers from the neighbouring countries – Guinea, Benin, Ivory Coast and Togo – also took part in the project to widen its potential impact.

Aquaculture

R6759 Integration of aquaculture into the farming systems of the eastern plateau of India (also Shujaaz)

The eastern plateau region of India is characterised by a short rainy season, limited water storage capacity and a prolonged dry season. Farming systems provide only one rice crop per year. Aquaculture is limited to extensive stocking of fry, mainly in perennial water bodies, with no management practised. The adoption of group-based aquaculture in the project area proved significant, with 57 per cent of all farm groups within the EIRFP conducting managed aquaculture by the end of the project. The concept of 'staged' fish production i.e. the production of various stages of fish locally (hatchlings, fry, fingerlings) apart from food fish was also trialled with groups. This demonstrated that decentralised fish seed production was possible and could meet a variety of needs. It also suggested that linkages with current private sector seed networks could be strengthened to benefit a wider range of stakeholders and improve the sustainability of the approach.

R8100 Investigating improved policy on aquaculture service provision to poor people (also Shujaaz)

Fish culture has a long tradition in India, although the development and documentation of options suited to poor people's objectives and resources is recent. A key feature of the project's overall process was to provide professionally facilitated 'space' for farmers, fishers, service providers, policy implementers and policy makers to express their views in a series of local and national meetings and workshops. Carefully facilitated, multilingual meetings and workshops were conducted to promote national, multilevel stakeholder discussions of modes and priorities for policy change. Understanding was built about the process of transacting policy and institutional change, highlighting lessons learnt from elsewhere, defining priorities for change and developing indicators which demonstrate progress towards change. The project engaged with policy-makers and implementers so as to stimulate debate and moves towards policy change. A semi-anonymous Consensus-building Process was used by policy actors to prioritise changes. A portfolio of 42 policy-change priorities was transacted to 13 top priority recommendations. Importantly, the process brought through the voices of poor people to a senior policy level.

R8334 OASIS: The One Stop Aqua Shop (also Shujaaz)

This project focused on promoting the findings and products of R8100 with national and state level stakeholders. Its purpose was to further develop and promote mechanisms for the delivery of improved rural aquaculture services at the state and national levels in the target eastern India states that were highlighted from previous DFID Natural Resources Systems Programme projects in Jharkhand, Orissa and West Bengal. It was envisaged that by project end, stakeholders in at least two state-level institutions and some key national-level policy actors would use new knowledge generated by the project in ways that can benefit people who are poor. Lessons were documented about a process of 'facilitated advocacy', to give farmers and fishers a voice in policy-making and a 'consensus-building process' (CBP) around change priorities within line agencies. The project worked with stakeholders to nurture and monitor change through communications and institutional capacity building. New communications vehicles developed by the project were translated and shared throughout the region. Policy briefs and better-practice guidelines as outputs from the project were also taken up by other projects and organisations.

R8363 Enhancing development impact of process tools piloted in eastern India (also Shujaaz)

The project aimed to improve the delivery of aquaculture support services to farmers and fishers by encouraging promotion and uptake of tools within government and non-government institutions, by STREAM staff and by partners in Asia-Pacific. The tools are the entry-point of encouraging and supporting self-help groups, the consensus-building process and information access surveys. The expectation was that stakeholders in Cambodia, Indonesia, Lao PDR, Nepal, the Philippines, Sri Lanka and Vietnam could learn about the process tools from eastern India and develop ways of sharing and using them in their own work. The process resulted in the production of two new genres of publications, (i) Better-Practice Guidelines – lively, colourful four-page media with cartoons, text and images aimed at those working closely with communities, and (ii) Policy Briefs – specialised two-page précis with links to other information. This led to the sharing of twelve country specific versions of each policy brief and better-practice guideline in local languages, which assisted in generating rapid visibility in Asia-Pacific.

Seed yam

R5259 An examination of Dioscorea spp (yam) for nematode resistance and its incorporation into improved cultivars

The project established and maintained a collection of yam species and breeding lines. Screening techniques for the assessment of resistance in yams to *S.bradys* and *P. coffeae* (involving the large scale culture of both nematodes) were developed. Good communication (and exchange of germplasm) has been established with yam breeders in Nigeria (IITA) and the project is now well placed to begin to provide information to breeders.

R5345 Epidemiology and control of yam anthracnose

Isolates of *C. gloeosporioides*, collected from *D. alata*, other species of Dioscoreas and alternative hosts were found to be highly variable in terms of their morphological and cultural characteristics, although this could not be related to origin of isolate or virulence of fungus on *D. alata*. Detached leaf experiments to determine pathogenicity of the isolates on *Dioscorea spp.* showed that pathogenicity varied according to the host plant from which isolate was collected and, to a lesser extent, geographic location. Isolates from some alternative hosts e.g. *Citrus sp.* and *Mangifera indica* were highly pathogenic on *D. alata* and are considered to be a potential source of inoculum. A field survey, carried out to investigate the temporal progress of yam anthracnose, showed disease development was influenced by rainfall and fungicide application. Lower rainfall and a regular fungicide spray programme initiated prior to symptom development delayed the onset of disease and resulted in a slower rate of disease increase. The cost of a spray programme is, however, prohibitive to the majority of yam growers who are resource-poor farmers, and the fungicides are not always effective. Cultural methods of control (intercropping and date of planting) were also found to be ineffective against anthracnose on susceptible varieties of *D. alata*. Tuber-borne inoculum is identified as the tuber source of inoculum initiating epidemics of anthracnose. The use of 'clean' planting material and identification of methods of cleaning contaminated seed-tubers are considered to be integral components in formulating a control strategy. The identification and cultivation of anthracnose-resistant varieties should increase yam production although dependence upon a single variety must be avoided since field experiments suggest that resistance is breaking down in some reportedly anthracnose-tolerant varieties. Further research into mechanism by which inoculum is transmitted from the tuber to foliage, and the durability of anthracnose-resistant varieties is recommended before effective control strategies can be implemented.

*R5346 Biology of yam anthracnose (*Colletotrichum*)*

The wider objective of this project is to identify the major diseases of West African crops and to develop control strategies through knowledge of their biology and ecology. The immediate objective is to develop a project on the epidemiology of yams in West Africa. Anthracnose caused by *Colletotrichum gloeosporioides* has been identified as a major constraint to yam production in West Africa where 80–90 per cent of the world yam production occurs. The project has provided improved knowledge of the important fungal pathogens of West African crops, improved knowledge of the fungal pathogens of yam, especially yam anthracnose and improved control strategies for yam anthracnose.

R5688 Natural resistance to infection by Colletotrichum

The objective of this project was to investigate the probability that resistance in yam and tomato plants is due to the presence of antifungal compounds and that the levels can be modulated to induce resistance/tolerance to infection. In particular, *Colletotrichum gloeosporioides* has been identified as a ubiquitous pathogen causing serious diseases on many crops. Certain cultivars of tomato and yam have been shown to exhibit resistance/tolerance to fungal attack and contain or produce larger quantities of antifungal compounds which may be involved in resistance. The work on yams was linked to projects in Barbados (CARDI) and Reading University (R5688/X0235). The tomato work was developed from a project (X0047) based at Nottingham University.

R5983 Factors influencing the occurrence of yam tuber rots in West Africa

Yam (*Dioscorea sp.*) is typically planted in heaps or ridges, and is often the first crop planted at the start of the cropping period. It has a relatively high demand for labour, largely because of the need for good land preparation, staking and care during harvesting. There are two main constraints in yam cultivation. Firstly, planting material (small tubers or cut pieces of tuber) is expensive and secondly, the harvest crop is prone to damage during storage and transportation. A method for producing planting material more cheaply by the use of minisetts has been developed by the International Institute for Tropical Agriculture (IITA) and the National Root Crops Research Institute (NRCRI), and this is being further developed by the NRI Adaptive Research Initiative project (F0006). However, there is still need for much more basic research, especially at the household level, into the causes of loss during storage. The project provided:

- Information on yam storage rots in Nigeria, linking causal agent with cultivation, harvest and storage practices.
- Identification of the fungal rot pathogen *Fusarium* (not *Botryodiplodia* as reported in the literature) as the prime agent.
- Simple diagnostic technologies for yam rot pathogens in Nigeria and staff trained in their use.
- Recommendations for appropriate strategies to minimise losses.
- Identification of the major storage insect pests.

R5738 Epidemiology and control of anthracnose disease of yam in Nigeria

The wider objective of this project was to develop an understanding of the epidemiology of yam anthracnose in West Africa. The immediate objectives were to survey anthracnose in Nigeria and relate to cropping practices, to examine the survival of anthracnose between seasons and to investigate the host range and variability of *Colletotrichum* on yam in Nigeria. It concluded that chemical control of the disease is expensive and inappropriate for the resource poor farmers of the region. In order to develop appropriate integrated disease control strategies it will be necessary to gain a better understanding of the epidemiology of the disease.

R6691 Control of yam diseases in forest margins farming systems in Ghana

The main objectives of the project were to determine the nature and impact of yam diseases in Ghana. Recommendations for improved and sustainable pest management practices were formulated and tested on-station. These recommendations were promoted to smallholder farmers. A rapid rural appraisal of farmers' perceptions of yam pests and diseases was carried out during January 1998. Pests and diseases, grouped together, were ranked the second most important problem (after lack of finance) and were considered a major reason for the poor yam yields experienced by the majority of farmers in 1997. Furthermore, all but one of the farmer groups questioned said that pests and diseases had been increasing in severity over the previous five years.

R6694 Identification of resistance to major nematode pests of yams in West Africa

This project sought to identify sources of resistance to the principal nematode pests of yams (*Scutellonema bradys* and *Meloidogyne incognita*) which cause dry rot and galling of yam tubers in West Africa and are a priority for yam breeders at IITA. The project focused on development of yam nematode resistance screening techniques for field and screenhouse and carried out extensive screening of germplasm from West Africa. An important aim was to achieve a better understanding of the genetic and environmental components of variability in the host: parasite interaction of yams and yam nematodes through strategic research. The project involved survey work, which examined the distribution of nematodes in Ghana and assessed farmer's perceptions of nematode losses. It also involved field, screenhouse and glasshouse trials in Nigeria, Ghana and the UK.

R8278 Evaluation and promotion of crop protection practices for 'clean' seed yam production systems in central Nigeria

This was primarily a promotional project with the main purposes of: a) evaluating crop protection practices (based on current practices, local technical knowledge and the outputs from previous projects) for clean seed yam production for applicability and economic efficacy in Kogi and Ekiti states of Nigeria, and b) developing dissemination outputs related to clean seed yam production which would lead to improved seed yam health and availability through promotion within Nigeria initially, but would also be suitable for wider applicability across the yam-growing belt of West Africa. The project has succeeded in establishing a loose network of stakeholders in Nigeria, who have a strong interest in improving seed yam production. This has resulted in the call for the establishment of a seed yam growers association in Nigeria, which would not only act to support growers and extension workers, but would also provide a platform for influencing regional and national policy related to yam production and seed supply.

R8416 Upscaling sustainable clean seed yam production systems for small-scale growers in Nigeria

Previous crop protection programme projects and others have shown that scarcity and expense of clean planting material (seed yams) is a major constraint to increasing yam production and productivity in West Africa. This project was set up to follow on from previous projects in West Africa. The main aims were to gain a better understanding of the systems and economics of seed yam production and supply, and to further evaluate, demonstrate and promote the systems developed in the earlier projects for producing good quality seed yams. Although treating the planting material before planting will increase the quality and storability of the seed yams, further research is required to find methods to maintain or restore soil fertility while still using the short fallow system, perhaps by inter-planting with a nitrogen-fixing legume, or making appropriate formulations of organic or inorganic fertiliser available at an affordable cost. There are three main systems of production within Nigeria where further interventions are likely to improve the general availability of good quality seed yams.

NERICA rice

R7251 Promotion of pro-poor strategies to reduce the impact of key pests and diseases in vegetable crops

The objective of this project was the promotion of pro-poor strategies to reduce the impact of key pests and diseases, improve yield and reduce pesticide hazards in peri-urban systems. It contributed to sustainable rural livelihoods in that the outputs help farmers to produce their vegetable crops (for consumption and sale) in a safe, more effective and economic way. Benefits include improved nutrition for whole families, better cash returns from higher yields of better quality produce and an empowerment through agricultural knowledge which help them to make informed choices on other cropping options.

R8312 Promotion of quality vegetable seed in Kenya (2003–05) (also Shujaaz)

Market gardening and horticultural enterprises represent a significant source of income for many small-to-medium scale growers in Kenya. Brassicas are an important part of the diet of Kenyans, especially in low-income groups. Kale and cabbage are among the most important crops grown by smallholders. A socio-economic survey identified the types of kale that farmers use, their preferences and the sources farmers use for seed. An inventory of brassica seed in Kenya has been drawn up from commercial seed companies/local markets. Farmers' views on current seed production and marketing systems that exist in Lari division were documented and the feasibility of a community-based seed production and seed marketing strategy explored. PRA activities have thus made significant contributions to our understanding of farmers' perceptions and needs with respect to seed purchases. Kinale farmers expressed a strong interest in multiplying/marketing seed with improved seed health and quality. In close collaboration with KEPHIS inspectors, using international UPOV guidelines, Kinale kale has been characterised as a variety for the first time.

R8439 Promotion of quality kale seed in Kenya (also Shujaaz)

The use of quality seeds along with other inputs and appropriate cultural management practices is recognised as the most cost effective way of increasing crop production and productivity. Seeds from the five best potential kale varieties identified (CABI 1-5) were submitted to KEPHIS, with documentation detailing their specific characteristics. CABI kales 1–5 were planted to evaluate their performance, compared to local varieties, in different agro ecological zones in peri-urban Nairobi and Western Kenya. Their performances consistently surpassed the commonly grown commercial variety. Seeds of all lines developed, including CABI 1–5, have been deposited in the KARI genetic resources unit and in the vegetable gene bank (Warwick-HRI, UK). On-farm participatory demonstration plots were established with existing farmer groups. In preparation for continuous multiplication and commercialisation of the improved kale seeds, Lari Division farmers had extensive discussions amongst themselves and the CDA. As a result, they obtained official registration and authority from the District Social Development Officer, under the National Community Development Programme in Kenya.

R8480 The Good Seed Initiative (GSI) – sharing the learning from programmes into pro-poor seed systems in East Africa

This project learned lessons and produced guidelines on farmer level production of quality seed (cf. the kale and cabbage seed projects noted above) which are directly applicable to this project. Farmer-saved and farmer-traded seed (i.e. the informal seed sector) is, and for the foreseeable future will continue to be, the major source of seed of staple crops in sub-Saharan Africa. Recognising the important and vital role played by the informal seed sector, CAB International (CABI) launched the 'Good Seed Initiative' (GSI) – a global initiative that seeks to strengthen the capacity of small-holders to source, produce, manage and disseminate seed, and thereby contribute to food security and improved livelihoods. Over the past 15 or so years, the DFID Crop Protection and Crop Post Harvest Programmes, together with other donors, supported projects which have generated seed-related research outputs on ways to produce and manage good seed, to promote new varieties and the *in situ* conservation of indigenous agro-biodiversity. By sharing the learning from these and other projects with the GSI, the project aimed to produce key dissemination outputs and provide an important dissemination pathway for the accelerated uptake of seed-related outputs by poor farmers in East Africa.

FIPS

R5237 Biology and management of termites and white grubs in smallholder cropping systems

This project identified all major pest termite species and quantified the damage they do to the important subsistence crops in representative agro-ecological zones throughout the southern region of Malawi. In particular, it established that termites can damage up to 25 per cent of maize crops (the staple cereal) and up to 30 per cent of cotton crops (a cash crop in the drier areas). It has also established a clear negative correlation between termite damage to crops and mean annual rainfall. This may be useful in decisions as regards resource allocation for termite control. Experiments on the effects of inter-cropping and residue management in termite damage to maize have also demonstrated that inter-cropping reduces termite damage but the reduction is often too small for inter-cropping to be effective on its own. There is considerable variation in damage according to season and location and the factors causing this variation need to be clarified. Neither inter-cropping nor soil management techniques appear to have a significant effect on white grub damage to maize. A number of pest white grub species have now been identified in the adult stage and a key for their identification provided.

R6642 Development of management strategies for maize streak virus disease – 1

This project concentrated on the development of management strategies for maize streak virus disease (MSVD) based on cultural control practices that are acceptable to farmers. These were developed through an understanding of the behaviour of the leafhopper vectors and their role in MSVD spread within maize plots and between maize plantings. The unreliability of available maize seed was a major constraint to farmers producing good yields from their maize crop. Through collaboration with village groups, extension officers and NGOs it has been possible to train farmers in the study villages and to provide them with small amounts of high quality Longe 1 breeders' seed. The project has begun exploring the idea of empowering farmers to produce their own seed through selection and controlled pollination. The farmers have been very enthusiastic about taking up this technique and a follow-up of this work has been agreed.

R7429 Development of management strategies for maize streak virus disease – 2

This project was an extension of the one outlined above. Streak-resistant cultivars provide an effective means of controlling maize streak virus disease (MSVD) in many farming systems in Africa. In Uganda, the National Cereals Programme has released the streak resistant cultivar Longe 1 for use by low-input subsistence farmers. The project completed the validation of a system whereby farmers were provided with knowledge of how to produce good quality seed of a superior MSV-resistant maize variety through researchers collaborating with an NGO and several community-based organisations and with locally developed training materials and farmers trained as trainers. The quality of the maize seed produced by the farmers was validated by on-station trials and the efficiency of the various training methods used was assessed by interviewing participants. A novel aspect of the training was to use a locally produced video shown through existing mobile video TV systems which normally show popular video films in the villages.

R7404 Socio-economic study of the uptake of herbicide technology in maize cropping systems

The purpose of this project was to provide a critical assessment of the net benefits to rural communities associated with herbicides, and to identify those factors constraining their adoption. Both of these goals were achieved by this, indicating clearly that herbicides could be expected to produce positive benefits for communities. However, their adoption is constrained by a combination of poverty, poor access to credit, temporal cash flow issues and an undervaluing of female labour time.

R8219 Improved access to appropriate farm inputs for integrated maize crop management by small-scale farmers in Kenya

This project demonstrated that there are three major bio-physical constraints to improved maize crop production by small farmers in Embu and Kirinyaga districts in Kenya: poor soil fertility, prevalence of the maize streak virus, and weeds. Use of appropriate farm inputs (fertilizers, disease-tolerant varieties and herbicides) is essential to increase yields. However, inputs are conventionally packaged in large bag sizes that are too expensive for most small farmers. The project introduced appropriate farm inputs and crop management, including Farmer Field Schools (FFS), small plot demonstrations (maize varieties, fertilisers and herbicides) and promotions using small packs of inputs provided by the private sector. In addition, FIPS-Africa conducted a total of 1,500 small plot demonstrations in six districts over four seasons. Project outputs were promoted through a network of 120 farm input agro dealers who were trained in the identification and control of MSV. With support from the private sector, the geographical range of the project was extended into Nyeri, Meru, Kiambu, Thika, Murang'a and Maragwa districts.

R8220 Improving farmers access to and management of disease resistant cultivars in the southern Highlands of Tanzania

This project sought to develop and promote strategies to reduce the impact of pests and improve the quality and yield from high potential maize cropping systems for the benefit of poor people in the Southern Highlands of Tanzania. Activities were aimed to realise three main outputs: (1) Disease resistant maize varieties appropriate to farmers' needs and adapted to local conditions validated by farmers and other stakeholders; (2) Approaches for improving access to and management of quality seed by farmers validated and promoted; and (3) Sustainable pathways/systems for quality seed supply appropriate to local conditions and farmers, needs developed by farmers and other stakeholders. The two most important yield-limiting maize diseases in the Southern Highlands and the country as a whole (Grey leaf spot and Maize streak virus) were addressed, using resistant cultivars identified during project activities as the main control strategy. The project facilitated the establishment of a public-private partnership between KARI and private sector seed companies.

R8455 Improved access to appropriate farm inputs for integrated maize crop management by small-scale farmers in Kenya and Tanzania

This project expanded the variety of inputs available to purchase by farmers and explored the opportunity to take this work forward also in Kenya. An impact assessment found increases in food security amongst farmers who had adopted the use of MSV-tolerant maize varieties and improved fertilisers promoted by FIPS-Africa in the preceding project. A subsequent survey of 100 farmers by FIPS-Africa showed that the number of food secure families increased from 30 to 80 per cent. The extra quantity of maize produced per farmer was on average 5.65 bags x 90 kg bags. A total of 1,012 demonstrations of MSV-tolerant varieties, and optimum fertiliser and weed management were conducted during this period.

Shujaaz

R5539 Commercialisation of solar drying technologies for micro and small-scale rural enterprise development

Poor farmers are avoiding the waste and low returns associated with overproduction of fruits and vegetables thanks to simple solar-drying techniques. Processing in this way helps preserve the quality of produce and provides opportunities for farmers to add value for local, regional and international markets. Enterprises known as primary marketing organisations (PMOs) are taking the lead in creating a commercially viable value chain, helping farmers to introduce the new technologies and access markets. In Uganda, more than 700 fruit farmers at 85 sites – mainly women – are using 110 solar dryers. The equipment, and the associated business model, are also being used by poor smallholders in a range of other developing countries, including Burkina Faso, Colombia, Ghana, India, Pakistan, Sri Lanka and Zambia.

R6619 Husbandry strategies for forages to increase milk production from cows and goats on smallholder farms in Tanzania

There was little evidence that the dairy production constraints facing poorer farmers in Tanzania had been adequately identified or addressed prior to this project. This project applied participatory appraisal techniques so that farmers could identify and prioritise their constraints and participatory evaluation techniques and evaluate experimental technologies for themselves. Finally, farmer-to-farmer learning and evaluation permitted the transfer of potentially beneficial technology from one group of farmers to another. The adoption of a participatory approach to all stages of the technology generation and dissemination cycle is rare in livestock research and this project serves as an example of the benefits which may accrue from this approach.

R7571 Management of virus diseases of vegetable crops in Kenya

Smallholder horticultural production is an important and expanding component of rural livelihoods in Kenya. Vegetable production provides employment and income for farmers, their families and employees. Pest and disease damage threatens the yield of crops. Vegetable farmers' usual response is heavy and frequent application of pesticides. These are expensive and are often unsuccessful in protecting crops, especially against virus disease. Reliance on pesticides has led to increasing concern about residues in produce, operator safety, pesticide resistance and environmental damage. Few alternatives are currently in use. A study confirmed high levels of virus in cabbage, cauliflower and kale crops, but the impact of these viruses on crop yield was unknown. The overall aim of the project was to develop improved methods for the control of virus diseases, in particular Cauliflower mosaic virus and Turnip mosaic virus, in brassica crops in the peri-urban vegetable systems being studied within the vegetable project cluster in Kenya. The project developed recommendations in several key areas for improved control of virus diseases.

SOS

R7364 Improving the control of tsetse: the use of DNA profiling to establish the feeding of tsetse to cattle

This project aimed to develop and promote environmentally beneficial and cost effective strategies for the control of trypanosomiasis in Zimbabwe and other tsetse-infested countries of sub-Saharan Africa. This work contributed towards developing a sustainable, environmentally beneficial and cost-effective strategy for controlling trypanosomiasis using insecticide-treated cattle. The work indicated that a selective dipping strategy, based on the treatment of the larger animals within a herd can achieve more cost-effective control thereby offering a cost-effective strategy for poorer livestock-keepers in Zimbabwe, Tanzania and Ethiopia who typically have small heterogeneous herds of cattle.

R7360 Field methods and tools for resource-poor farmers and extension workers to improve targeting and appropriate use of drugs used to control African bovine trypanosomiasis

The supply and value of animal products and contribution of livestock to crop production is severely compromised in tsetse-infested areas of Africa through the effects of bovine trypanosomiasis. Trypanosomiasis reduces cattle density by 37–70 per cent, and the offtake of meat and milk by about 50 per cent. Trypanocidal drugs, among the few animal health interventions available, are misused and overused with resultant low benefit/cost ratios, and increasing risk of drug resistance. These trypanocidal drugs are the only means of controlling bovine trypanosomosis that are widely available to smallholder farmers, who spend $30 million p.a. on these drugs. Surveys have shown that farmers lack the knowledge to use them appropriately. The project provided improved methods of chemotherapy by target stakeholder groups, including smallholder farmers, extension workers, women's groups and school children; and robust delivery media and formats appropriate for education of all stakeholders in these improved methods.

R7538 Review of environmental change and sustainable, poverty-focused strategies for trypanosomiasis control in Africa

Trypanosomiasis has long been recognised as a major constraint on animal husbandry, livestock production, and mixed farming in tropical Africa. This review showed greatly improved understanding, but questioned how this knowledge was to be applied at farmer level. It recommended further support for adaptive and well-focused strategic research, targeted at three specific areas of interest: (a) identification of priority areas for disease control; (b) assessment of disease risk and control impacts and (c) integration of disease control and agricultural development. All three topics were addressed to provide a retrospective assessment of trypanosomiasis control in the context of agricultural expansion and environmental change across Africa, and identify sustainable, poverty-focused disease control strategies for the future.

R7596 Decision support system for the control of trypanosomiasis in south-east Uganda: improving public health and livestock productivity through cost-effective control of trypanosomiasis in livestock

In SE Uganda trypanosomiasis constitutes a major health problem that affects both man and his livestock. Cattle are the primary reservoir of Trypanosoma *brucei rhodesiense*, the causative agent of *rhodesiense* sleeping sickness. Sleeping sickness is fatal if not treated and control of the disease has traditionally relied on tsetse control and active case detection and treatment – expensive activities dependent on public funding and a high degree of community participation to ensure sustained impact. An extension in the geographic distribution of sleeping sickness to the north of the traditional disease focus in SE Uganda was linked to movement of cattle as part of a restocking programme in Teso. Thus, the role of cattle as a reservoir of sleeping sickness might further constrain rural development through restricting trade, movement and restocking of cattle. The project helped resolve the role of cattle in the maintenance and spread of human sleeping sickness in SE Uganda, and the lack of cost-effective and sustainable control options for animal trypanosomiasis and human sleeping sickness. The project enabled policy makers, public institutions, communities and individual smallholder farmers to identify appropriate and cost effective methods for the sustainable control of trypanosomiasis in the cattle of SE Uganda, through development of cost-effective and appropriate strategies to sustainably control diseases of livestock that affect the livelihoods of the poor.

R7597 Development of a low-cost haemoglobinometer and other diagnostic tools for bovine disease diagnosis in sub-Saharan Africa

The project provided for the validation of rapid, reliable and cheap diagnostic tests for bovine disease control, and associated user-friendly, field level decision support tools. These were a low cost, pen-side method of measuring haemoglobin (a measure of anaemia) and a supplementary low-cost, low-technology decision support tool. Several candidate systems for measuring haemoglobin were identified and evaluated under laboratory and field conditions. These ranged from simple colour charts for use with filter paper strips to hand-held colorimetric haemoglobinometers. When testing in cross-sectional and longitudinal field studies of cattle under local conditions in smallholder mixed crop-livestock production systems in East Africa, a hand-held haemoglobinometer was found to be a simple and effective means of determining haemoglobin levels. These same studies were used to provide data on the prevalence, incidence and impact of a number of common endemic bovine diseases including trypanosomiasis, theileriosis, anaplasmosis, babesiosis, cowdriosis, haemonchosis, schistosomiasis and fasciolosis. Finally, a Delphi survey of expert opinion of animal health-field veterinarians and veterinary scientists was conducted to obtain quantitative information of key clinical signs for endemic bovine diseases. The results of the Delphi survey were incorporated into a low-technology decision support card suitable for use by farmers and livestock keepers themselves, community animal health workers, animal health and veterinary assistants, other extension workers and qualified veterinarians.

R7539 Environmental risks of insecticide-treated cattle in semi-arid Africa livestock systems

This project sought to understand the transport and fate of insecticide residues in cattle dung, the ecological significance of residues for dung fauna and dung dispersal, and to offer mitigation measures that reduce any negative environmental impacts. Pyrethroids applied to cattle for the control of tsetse fly can contaminate dung sufficiently to affect fauna utilising dung as a resource. Reduced abundance of insects, dispersal of dung and productivity of pasture are potential impacts of insecticide treated cattle. Predictive population modelling indicated that when cattle treatments occurred over wide areas and for many months, the effects of residues on the abundance and distribution of dung fauna could be serious, threatening the slow breeding species (large dung beetles) and cattle frequenting muscoids. The route of contamination from treatment to dung was elucidated using mostly pour-on and dip formulations of *deltamethrin* (Spot-on and Decatix respectively) and employing the bioassay to ascertain the levels of contamination in the dung produced. Restricted applications reduce risks to dung fauna without compromising efficacy to tsetse. Significant cost savings (up to 90 per cent) would accrue from reduced insecticide use, and the switch from expensive pour-on formulations to water-based dip and spray products.

R7987 Message in a bottle: disseminating tsetse control technologies

The purpose of this project was to validate, promote and disseminate strategies to improve sustainably the health and productivity of livestock maintained by poor livestock keepers in semi-arid production systems. Work showed that by applying insecticide to leg and belly regions at 2–3 week intervals, the cost of the technique can be reduced by ~90 per cent, allowing farmers to control tsetse for <US$1/animal/year. Studies on the efficacy using wood smoke showed that smoke halved the biting rate, and that barriers constructed from netting or wood reduced the feeding rate by ~90 per cent. The results suggest that for owners of zero grazed cattle, a simple physical barrier would reduce the biting rate by >90 per cent which, in areas of low to moderate challenge (<1 bite/animal/day), would reduce the impact of trypanosomiasis significantly. A computer-based system to help NGOs design and implement community-based interventions against tsetse was developed. The system includes an interactive programme ('Tsetse Plan') to help users design and implement community-based interventions against tsetse using bait technologies. The programme synthesised knowledge from 20 years of research and, using a user-friendly interface, allowed non-specialists to assess the feasibility, cost and impact of various tsetse control strategies.

R8214 Integrated vector management: controlling malaria and trypanosomiasis with insecticide-treated cattle

In Africa, there are 250–450 million cases of malaria per year, and the annual economic losses associated with this are estimated to be ~ $US 2 billion. Some 10 million km^2 of the malaria-affected regions of sub-Saharan Africa are also infested with tsetse flies which transmit trypanosomiasis, a complex of diseases affecting the health and productivity of humans and livestock, which has been estimated to cost Africa $US 4.5 billion a year. The use of insecticide-treated cattle to control tsetse is an increasingly important means of controlling trypanosomiasis, especially for poorer livestock owners. In the more arid, tsetse-affected regions of east and southern Africa, the main vector of malaria is *Anopheles arabiensis*, which obtains a significant proportion of its bloodmeals from cattle. Theoretical evidence suggests that in areas where malaria is transmitted by this vector, the insecticidal treatment of cattle could reduce malaria incidence. The project showed that while the insecticide formulations currently used to treat cattle to control tsetse and ticks will not have a significant impact on malaria transmission, if cattle were treated with a formulation effective against tsetse and *An. arabiensis*, then insecticide-treated cattle could make a significant contribution to the control of malaria in the areas where *An. arabiensis* is the main vector. The project also developed a rational framework for assessing the likely impact of insecticide-treated cattle on malaria vectors.

R8318 Decision-support for endemic diseases in sub-Saharan Africa, private sector drivers for technology adoption by poor livestock keepers

Endemic diseases of sub-Saharan African cattle are a major constraint to sustainable rural livelihoods but their control is constrained by contraction of veterinary services, with devolution of diagnosis and treatment to less well trained cadres of veterinary service providers, change agents and non-professional cattle keepers who lack clinical knowledge or diagnostic tools. Anaemia is a cardinal sign of many important vector-borne and other endemic diseases of cattle, particularly in the smallholder rainfed humid farming system. Anaemia or its absence has been shown to be an excellent indicator of bovine health in vector-borne or helminth disease endemic areas. A novel, low-cost haemoglobinometer developed in conjunction with a commercial partner, enables animal health workers to measure anaemia in domestic livestock. The instrument is small enough to be carried on foot, on a bicycle or motor scooter (typical modes of transport for end-users), uses a sampling and measuring consumable costing a few pence, and is battery operated or solar powered. The decision support tool was developed as a diagnostic method complementary to the haemoglobinometer, and focuses on eight endemic bovine diseases, including trypanosomiasis, helminths and tick-borne diseases, which are widely recognised as key constraints to animal productivity. The tool is a colour-banded card utilising a scoring system and differential diagnosis is performed by comparing observed clinical signs (including anaemia) with disease profiles and constructing a ranked list indicating the most likely diagnoses.

R8461 General model for ITC restricted application of insecticide to cattle

Tsetse-borne trypanosomiasis is a severe constraint to the livelihoods of poor livestock owners across ~10 million square kilometres of sub-Saharan Africa. The effects on cattle mortality, reproductive performance and productivity reduce ownership of, and access to natural, financial and human capital, and increase the vulnerability of many livestock keepers, including pastoralists, mixed crop-livestock and landless livestock keepers. Effective control of trypanosomiasis by the poorest livestock keepers is constrained by the relatively high cost of tsetse control and a general lack of access to appropriate information on control methods and strategies.

The cheapest method of controlling tsetse is the use of insecticide-treated cattle. This project was aimed at reducing the cost of this method by identifying means of reducing the amount of insecticide used, thereby reducing costs and impact on non-target fauna while still controlling tsetse effectively.

R8459 Tsetse Muse: an interactive programme to assess the impact of control operations on tsetse populations

Despite many models being available they are generally used only by the persons who produced them and the outputs are generally confined to specialist scientific literature. Tsetse Plan, an interactive computer programme was developed under a previous DFID activity allowing some models to be readily useable by non-specialists, and are generalised to offer wider ranges of inputs and outputs. This project developed a user-friendly decision-support tool called 'Tsetse Muse'.

Bibliography

Adolph, B. (2010) *Rural Advisory Services Worldwide: A Synthesis of Actors and Issues*, paper prepared for the Global Forum for Agricultural Advisory Services, Triple Line Consulting, London, 22 October.

Ampofo, J. K. O., Kumar, P. L. and Seal, S. E. (2010) 'Integrated crop management for sustainable yam production', in *Yam Research for Development in West Africa – Working Papers IITA-BMGF* Consultation Documents, IITA, pp. 46–80.

Anderson, J. R. (2008) 'Agricultural advisory services', background paper for the World Bank *Development Report,* World Bank, Washington DC.

Anderson, J. R. (1991) 'The origins of the International Rice Research Institute', *Minerva 29* (1): 61–89.

Anderson, J. R., Levy, E. and Morrison, B. (1991) *Rice Science and Development Politics: IRRI's Strategies and Asian Diversity 1950–1980,* Clarendon Press, Oxford.

Bell, R. M. (2006) 'Science and technology for development', *Background Discussion Paper for the L20 Workshop*, UNU-MERIT, Maastricht, Netherlands, 7–8 March; http://www.l20.org/publications/21_sY_Sci_background_Bell.pdf.

Bell, R. M. (2007) 'Technological learning and the development of production and innovative capacities in the industry and infrastructure sectors of the least developed countries: what roles for ODA?' *Background Paper No. 10 for the Least Developed Countries Report*; UNCTAD, Geneva; http://archive.unctad.org/sections/ldc_dir/docs/ldcr2007_Bell_en.pdf.

Biggs, S. D. (1990) 'A multiple source of innovation model of agricultural research and technology promotion', *World Development 18*(11): 1481–1499.

Biggs, S. and Clay, E. (1981) 'Sources of innovation in agricultural technology', *World Development* 9(4): 321–336.

Biggs, S. D. and Smith, G. (1998) 'Beyond methodologies: coalition – building for participatory technology development', *World Development 26* (2): 239–248.

Bolo, M. O. (2012) *Learning and Innovation in Agro Export Industries,* Lambert Academic Publishing, Saarbrucken, Germany.

Bud, R. (2010) 'From applied microbiology to biotechnology: science, medicine and industrial renewal', *Notes Rec. R. Soc. 64,* Suppl 1: S17–S29 (September).

CGIAR (1992) *Annual Report*, CGIAR, Washington, D.C.

CGIAR (2010) *Who We Are,* http://www.cgiar.org/who/structure/system/.

Chambers, R. and Ghildyal, B. P. (1985) 'Agricultural research for resource-poor farmers: the farmer first and last model', *Agricultural Administration 20*: 1–30.

Cherry, A. (2005) 'Public–private partnerships to develop viruses as bio insecticides for Ghana and Benin', in A. Sweetmore, F. Kimmins and P. Silverside (2006) *Perspectives on Pests II: Achievements of Research under UK Department for International Development Crop Protection Programme 2000–05*. Natural Resources International Ltd, Aylesford, UK.

Clark, N. (1985) *The Political Economy of Science and Technology*, Blackwell, Oxford.

Clark, N. (1987) 'Similarities and differences between scientific and technological paradigms', *Futures 19*(1): 26–42.

Clark, N. G. (2002) 'Innovation systems, institutional change and the new knowledge market: implication for third world development', *Economics of Innovation and New Technology 11*(1 & 2): 353–368.

Clark, N. and Juma, C. (2013, forthcoming) *Long Run Economics*, Bloomsbury Academic, London.

Clark, N., Frost, A., Maudlin, I., Seward, P., Wainwright, H. and Ward, A. (2011) 'Putting research into use: a market failure approach', *International Journal of Technology Management & Sustainable Development 10*(3): 185–200.

Clark, N., Mugabe, J. and Smith, J. (2007) *Biotechnology Policy in Africa*, Nairobi, ACTS Press.

Clark, N. G., Stokes, K. and Mugabe, J. (2002) 'Biotechnology and development: threats and promises for the 21st century', *Futures 34*(9–10): 785–806.

Cleaver, H. (1972) 'The contradictions of the Green Revolution', *The American Economic Review 72*: 177–88.

Collier, P. (2008) 'The politics of hunger', *Foreign Affairs* 87(6): 67–80.

Conway, G. R. (1985) 'Agro ecosystem analysis', *Agricultural Administration 20*: 31–55.

Conway, G. (1997) *The Doubly Green Revolution*, Penguin, London.

Conway, G. (2012) *One Billion Hungry: Can We Feed the World?* Cornell UP, Ithaca.

Conway, G. and Waage, J. (2010) *Science and Innovation for Development*, UKCDS, London.

Cullather, N. (2004) 'Miracles of modernization: the Green Revolution and the apotheosis of technology', *Diplomatic History 28*(2): 227–54.

DFID (2004) *The Use of Science in UK International Development Policy*, House of Commons Science and Technology Committee, Thirteenth Report of Session 2003–2004, The Stationery Office, 26 October.

DFID Research Funding Framework (2005–2007). See www.DFID.gov.uk.

DFID (2005) *Growth and Poverty Reduction: The Role of Agriculture*, DFID policy paper, PD 044, December.

DFID (2006) *White Paper on International Development Eliminating World Poverty – Making Governance Work for the Poor*, CM 6876, July.

DFID (2009) *Eliminating World Poverty: Building our Common Future*, CM 7656, July.

Ekboir, J. (2009) 'The CGIAR at a crossroads: assessing the role of international agricultural research in poverty alleviation from an innovation systems perspective', *ILAC Working Paper No. 9*.

Esteva, G. (1996) 'Hosting the otherness of the other: the case of the Green Revolution' in F. Appfel-Marglin and S. Marglin (eds), *Decolonizing Knowledge: from Development to Dialogue,* Oxford University Press, Oxford, pp. 249–278.

Eyitayo, O. A., Anthony, T. O. and Theresas, I. (2011) Economics of seed yam production using Minisett technique in Oyo State, Nigeria. *Field Actions Science Reports, The Journal of Field Actions 4.*

FAO (2011) *World Aquaculture 2010,* Proceedings of the Global Conference on Aquaculture Phuket Thailand Technical Paper No 500/1. FAO Fisheries and Aquaculture Department, Rome.

Feder, G., Anderson, J. R., Birner, R. and Deininger, K. (2010) *Promises and Realities of Community-Based Agricultural Extension,* IFPRI Discussion Paper 00959, March, Washington DC.

Fitzgerald, D. (1996) 'Exporting American agriculture: the Rockefeller Foundation in Mexico, 1943–1953', *Social Studies of Science 16*(4): 457–483.

Fransman, M. (1995) *Japan's Computer and Communications Industry,* Clarendon Press, Oxford.

Georgiadas, N. J. and McNaughton, S. J. (1988) 'Interactions between grazers and a cyanogenic grass, *Cynodon plectostachyus*', *Oikos 53*(3): 343–350.

Gibbons, M., Limoges, C., Nowotny, H., Schwartzman, S., Scott, P. and Trow, M. (eds) (1994) *The New Production of Knowledge,* Sage, London.

Gildemacher, P. and Mur, R. (2012) *Bringing new ideas into practice: experiments with agricultural innovation. Learning from Research Into Use in Africa.* KIT Publishers, Amsterdam. 184 pages, ISBN 978-94-6022-233-7

Gildemacher, P. R., Leeuwis, C., Demo, P., Borus, D., Schulte-Geldermann, E., Kinyae, P., Mundia, P., Nyongesa, M. and Struik, P.C. (2012) 'Dissecting a successful research-led innovation process: the case of positive selection in seed potato production in Kenya', *IJTMSD 11*(1): 67–92.

Glaeser, B. (1987) *The Green Revolution Revisited: Critiques and Alternatives,* Allen and Unwin, London.

Grzywacz, D. and Rabindra, R. J. (2001) 'Use of insect viruses to control Helicoverpa', in A. Sweetmore, G. Rothschild and S. Eden-Green (eds), *Perspectives on Pests,* NR International Ltd, Chatham, UK, pp. 91–93.

Grzywacz, D., Cherry, A.C. and Gwynn, R. (2009) 'Bio pesticides for Africa: a case study on how research can better benefit Africa's poor', *Appropriate Technology 36*: 61–63.

Gummett, P. (1980) *Scientists in Whitehall,* Manchester, Manchester UP.

Hall, A. J. (2006) "Public-private sector partnerships in a system of agricultural innovation: concepts and challenges', *International Journal of Technology Management and Sustainable Development 5*(1): 3–20.

Hall, A. (2009) 'Challenges to strengthening agricultural innovation systems: where do we go from here?' in I. Scoones and J. Thompson (eds), *Farmer First Revisited,* Practical Action, Rugby.

Hall, A. J. and Yoganand, B. (2004) 'New institutional arrangements in agricultural research and development in Africa: concepts and case studies', in A. J. Hall, B. Yoganand, V. Rasheed Sulaiman, R. Raina, S. Prasad, G. Naik and N. G Clark (eds), *Innovations in Innovation: Reflections on Partnership and Learning,* ICRISAT, Patancheru, India and NCAP New Delhi, India.

Hall, A. and Nahdy, S. (1999) 'New methods and old institutions: the "systems context" of farmer participatory research in national agricultural research systems. The case of Uganda', *Overseas Development Institute (AgRen), Agricultural Research & Extension Network, Network Chapter 93*, January.

Hall, A. J., Yoganand, B., Rasheed Sulaiman V., Raina, R., Prasad, S., Naik, G. and Clark N. G. (eds) (2004) *Innovations in Innovation: Reflections on Partnership and Learning*, ICRISAT, Patancheru, India and NCAP New Delhi, India.

Hargrove, J. W., Ouifki, R., Kajunguri, D., Vale, G. A. and Torr, S. J. (2012) 'Modelling the control of trypanosomiasis using trypanocides or insecticide-treated livestock', *Public Library of Science Neglected Tropical Diseases* 6(5): e1615.

Hazell, P., Poulton, C., Wiggins, S. and Dorward, A. (2010), 'The future of small farms: trajectories and policy priorities', *World Development 38*(10): 1349–1361.

Hirvonen, M. (2011) *NERICA Seed versus Local Landraces: Another Battle of the Paradigms?*, RIU Discussion Paper 21, April.

Hornberger, K., Ndiritu, N., Ponce-Brito, L., Tashu, M. and Watt, T. (2007) *Kenya's Cut Flower Structure,* http://www.isc.hbs.edu/pdf/Student.../Kenya_Cut-FlowerCluster_2007.pdf.

Horton, D., Akello, B., Aliguma, L., Bernet, T., Devaux, A., Lemaga, B., Magala, D., Mayanja, S., Sekitto, I., Thiele, G. and Velasco, C. (2010) 'Developing capacity for agricultural market chain innovation: experience with the "PMCA" in Uganda', *JID* 22(3): 367–389.

IDL Group (2011) *Research Into Use: An Independent Review. Annex 9 Institutional Evaluation of RIU's Modalities and Experiments,* http://www.dfid.gov.uk/r4d/pdf/outputs/ResearchIntoUse/Research_into_Use_Independent_Review_Complete_Annexes.pdf.

Jamu, D. M. and Ayinla, O. A. (2003) 'Potential for the development of aquaculture in Africa', *NAGA, World Fish Centre Quarterly* 26(3): 9–13.

Juma, C. and Yee-Cheong, L. (2005) *Innovation: Applying Knowledge in Development.* Report of Task Force 10 on Science, Technology and Innovation, United Nations Millennium Development Goals Project, Earthscan, London, Sterling Va.

Juma, C. and Serageldin, I. (2007) 'Freedom to innovate: biotechnology in Africa's development', *African Union (AU) and New Partnership for Africa's Development (NEPAD) 5*(6).

Juma, C. (2011) *The New Harvest: Agricultural Innovation in Africa,* Oxford University Press, Oxford.

Kaplinsky, R., Chataway, J., Clark, N., Hanlin, R., Kale, D., Muraguri, L. and Wamae, W. (2009) 'Below the radar: what does innovation in emerging economies have to offer other low-income economies?', *International Journal of Technology Management & Sustainable Development* 8(3): 177–197.

Kenyon, L., Coyne, D., Claudius-Cole, B., Morse, S., McNamara, N. and Acholo, M. (2005) 'Evaluation and promotion of clean seed yam-production systems in Central Nigeria', in A. Sweetmore, F. Kimmins and P. Silverside (eds) (2006) *Perspectives on Pests II: Achievements of Research under UK Department for International Development Crop Protection Programme 2000–05*. Natural Resources International Ltd, Aylesford, UK.

Kuhn, T. S. (1970) *The Structure of Scientific Revolutions*, Chicago: University of Chicago Press, 2nd edn.

Lakatos, I. and Musgrave, A. (eds) (1972) *Criticism and the Growth of Knowledge,* Cambridge University Press, Cambridge.

Letty, B., Shezi, Z. and Mudhara, M. (2012) 'An exploration of agricultural grassroots innovation in South Africa and implications for innovation indicator development', *UNU-MERIT Working Papers 23*, March.

Lyall, C., Bruce, A., Tait, J. and Meagher, L. (2011) *Interdisciplinary Research Journeys: Practical Strategies for Capturing Creativity,* Bloomsbury Academic, London.

Machlup, F. (1962) *The Production and Distribution of Knowledge in the US,* Princeton University Press, Princeton.

Manyong, V. M., Alene, A. D., Olanrewaju, A., Ayedun, B., Rweyendela, V., Wesonga, A. S., Omanya, G., Mignouna, H. D. and Bokanga, M. (2007) *Baseline Study of Striga Control using IR Maize in Western Kenya* AATF/IITA Striga Control Project, IITA, Nairobi, September. See also www. aatf-africa. org/userfiles/IRmaizestudy.pdf.

Marglin, S. (1996) 'Farmers, seedsmen and scientists: systems of agriculture and systems of knowledge', in F. Appfel-Marglin and S. Marglin (eds), *Decolonizing Knowledge: from Development to Dialogue*, Oxford University Press, Oxford, pp. 185–248.

Maudlin, I. (2006) 'African Trypanosomiasis', *Annals of Tropical Medicine and Parasitology* 11(8): 679–701.

May, R. (2004) *Presidential Address to the Royal Society of London*, London, UK, December.

Morton, J. (2010) *The Innovation Trajectory of Sleeping Sickness Control in Uganda: Research Knowledge in its Context*, RIU Discussion Paper 08, October.

Mugagwa, J. T., Wamae, W. and Outram, S. M. (2010) 'Agricultural innovation and food security in Sub-Saharan Africa; tracing connections and missing links', *JID 22*(3): 283–289.

Muraguri, L. (2010) 'Unplugged! An analysis of agricultural biotechnology PPPs in Kenya', *JID 22*(3): 298–307.

Nicholas, H. G. (1971) *The United Nations as a Political Institution*. Oxford University Press, Oxford, 4th edn.

Nijihia, C. M. (2010) 'Shujaaz audience assessment case studies' *Synovate*, Nairobi, November 26.

Nowotny, H., Scott, P. and Gibbons, M. (2003) 'Mode 2 re-visited: the new production of knowledge', *Minerva 41:* 179–194.

Okello, A. (2012) *Beyond Avian Influenza: Policy Considerations for the Implementation of 'One Health' in Developing Countries,* PhD Thesis, University of Edinburgh.

Okello, A. (2013) *Beyond Avian Influenza: Policy Considerations for the Implementation of "One Health" in Developing Countries,* PhD Thesis, University of Edinburgh.

Pardey, P. G., Roseboom, J. and Beintema, N. M. (1997) 'Investments in African agricultural research', *World Development 25*(3): 409–423.

Perkins, J. (1997) *Geopolitics and the Green Revolution: Wheat, Genes and the Cold War,* Oxford University Press, New York NY.

Pray, C. (2002) 'The growing role of the private sector in agricultural research', in D. Byerlee and R. Echeverria (eds), *Agricultural Research Policy in an Era of Privatisation*, CABI Publishing, Oxon.

Riches, C., Lamboll, R. I. and Mbwaga, A. M. (2001) 'Integrated control of striga in Tanzania' in A. Sweetmore, G. Rothschild and S. J. Eden-Green (eds), *Perspectives on Pests: Achievements of Research under the UK Department for International Development's Crop Protection Programme, 1996–2000*. Natural Resources International Limited, Chatham, UK.

Rohrbach, D. D. (2010) 'Demand-led technological change: are public research and extension programs still relevant?' *Agreko 38*: S1, 156–164.

Rose, D. J. W., Dewhurst, C. F. and Page, W. W. (2000) *The African Armyworm Handbook; The Status, Biology, Ecology, Epidemiology and Management of Spodoptera exempta (Lepidoptera: Noctuidae)*. 2nd edn, Natural Resources Institute, Chatham, UK.

Roseboom, J. and Ruttan, H. (1998) 'The transformation of the Dutch agricultural research system: an unfinished agenda', *World Development* 26(6): 1113–1126.

Rothschild, Lord (1971) 'The organisation and management of government R & D', *Framework for Government Research and Development,* Cmnd 4814, HMSO, London.

Scoones, I. and Thompson, J. (2009) *Farmer First Revisited*, Practical Action, Rugby.

Sharp, M. (1989) 'The management and coordination of biotechnology in the UK 1980–88', *Phil. Trans. R. Soc. B 324*: 509–523.

Sharp, M. (1996) 'Biotechnology and advanced technological infrastructure policies: the example of the UK's protein engineering club', in M. Teubal *et al.* (eds), *Technological Infrastructure Policy: An International Perspective,* Kluwer Academic Publishers, Netherlands, pp. 217–246.

Singh, J. (1966) *Information Theory, Language and Cybernetics,* Constable, London.

Smith, J. (2009) *Science and Technology for Development*, Zed Books, London.

Snow, C. P. (1961) *Science and Government*, Harvard University Press, Harvard.

Solow, R. (1957) 'Technical change and the aggregate production function', *Review of Economics and Statistics 39:* 312–320.

SFSA (2011) *Syngenta Foundation for Sustainable Agriculture Review 2010–2011,* Basel, Switzerland.

Sweetmore, A., Kimmins, F. and Silverside, P. (2006) *Perspectives on Pests II: Achievements of Research under UK Department for International Development Crop Protection Programme 2000–05*. Natural Resources International Ltd., Aylesford, UK.

Sweetmore, A., Rothschild, G. and Eden-Green, S. J. (eds) (2001) *Perspectives on Pests: Achievements of Research under the UK Department for International Development's Crop Protection Programme 1996–2000*, Natural Resources International Limited, Chatham, UK. See also http://www.dfid.gov.uk/r4d/PDF/Outputs/CropProtection/CPPPerspectives.pdf.

Thirkell, A. and Wyeth, J. (2012) *Putting Innovation into Farmer Hands: What Works?* Document RIU, www.researchintouse.com.

Tripp, R. (1989). 'Farmer participatory research: new directions, old problems?' *IDS Discussion Chapter No. 356*, Institute of Development Studies, Brighton, Sussex.

Wabule, M. N., Ngaruiya, P. N., Kimmins, F. K. and Silverside, P. J. (2004) *Registration for Bio Control Agents in Kenya*, Proceedings of the PCPB/KARI/DFID CPP Workshop, Nakuru, Kenya, 14–16 May 2003. Kenya Agricultural Research Institute, Pest Control Products Board, Natural Resources International Limited. See also http://www.dfid.gov.uk/r4d/PDF/Outputs/CropProtection/R7299Registration.pdf.

Wambugu, F. and Kiome, R. (2001) *The Benefits of Biotechnology for Small Scale farmers in Kenya,* ISAA Briefs No 22 ISAA, NY.

Welburn, S., Coleman, P., Maudlin, I., Fevre, E., Odiit, M. and Eisler, M. (2006) 'Crisis, what crisis? Control of Rhodesian sleeping sickness', *Trends in Parasitology* 22(3): 124–128.

Wiggins, S., Kirsten, J. and Llambí, L. (2010) 'The future of small farms', *World Development 38*(10): 1341–1348.

Wilson, K., Grzywacz, D. and Mushobozi, M. (2009) 'Biological control of Armyworm in Africa: pitfalls and solutions', *Pesticide News 84*: 18–19.

Wiskerke, J. D. and van der Ploeg, J. D. (eds) (2004) *Seeds of Transition,* Royal van Gorcum, Assen, Netherlands.

Wooding, S., Hanney, S., Pollitt, A., Buxton, M. and Grant, J. (2011) *Project Retrosight: Understanding The Returns From Cardiovascular And Stroke Research*, Rand, Cambridge, UK.

World Bank (2006a) *Enhancing Agricultural Innovation: How to Go Beyond the Strengthening of Research Systems*, World Bank, Washington DC. See also http://siteresources.worldbank.org/INTARD/Resources/EnhancingAg_Innovation.pdf.

World Bank (2006b) *Meeting the Promise and Challenge of Sustainable Aquaculture*, World Bank, Washington DC.

World Bank (2007) *Changing The Face Of The Waters: The Promise and Challenge of Sustainable Aquaculture*, World Bank, Washington DC.

World Bank (2008) Development Report, World Bank, Washington DC.

Index